The Rocks at the Edge of the Empire

50 extraordinary rocky places that tell the story of the Romans and the landscape of their northern frontier

Ian Jackson

'Knowledge may have its purposes,
but guessing is always
more fun than knowing'

WH Auden 1973

First published in the United Kingdom in 2024 by Northern Heritage Services Limited

Units 7&8 New Kennels, Blagdon Estate, Seaton Burn, Newcastle upon Tyne NE13 6DB

Telephone: 01670 789 940 www.northern-heritage.co.uk

See www.northern-heritage.co.uk for full catalogue

Edited by Abigail Burt and Tony Cousins

Archaeological content reviewed by Nick Hodgson and Humphrey Welfare

Graphic design: Abigail Burt

Printed and bound in the UK by W & G Baird

British Library Cataloguing in Publishing Data

A catalogue record for this book is available from the British Library

ISBN 9781739486143

Society of Antiquaries

NORTHERN HERITAGE

WORLD LAND TRUST™
www.carbonbalancedpaper.com
CBP027042

FSC www.fsc.org MIX Paper | Supporting responsible forestry FSC® C016201

The old saying has it that 'geography is about maps and history is about chaps'. In this book, however, Ian Jackson skilfully weaves the two together to produce an informative insight into Britain's premier Roman monument, Hadrian's Wall. The two strands could not be more different when it comes to dates. The element of geography described in this book is geology, millions of years old, while the history is actually archaeology studying remains less than 2000 years old. Nevertheless, Ian makes a powerful case for studying the two together. The geology of northern England aids understanding of the very location of the Roman frontier and its various elements. But more than that, geology helps understand the fabric of the Wall, its forts, milecastles, turrets and artefacts.

This is an indispensable guide for all wishing to understand Hadrian's Wall better and the landscape in which it sits. Everyone who visits Hadrian's Wall after reading this book will see both through different - more informed - eyes.

David Breeze

Sycamore Gap
(Cover photograph: the Whin Sill east of King Arthur's Well)

Foreword

We warmly welcome this book which ranges across the interface between geology and archaeology. These two domains have many crossovers and yet the territory between them is not as frequently explored as it merits.

During my research on prehistoric querns from Yorkshire, I had the great good fortune to work with a colleague of Ian's from the British Geological Survey and I can testify to the immense value that the study of the geological environment can have for our understanding of the subsistence economy and industrial activity of past cultures.

This book provides a broad and accessible synopsis of the overlap. It is not a deep and detailed dive into any one aspect, but it nevertheless poses many questions that will provoke debate and that can also perhaps provide stimuli for further research and investigation. In that regard, it stands centrally within the long tradition of multidisciplinary research, often focussing on our spectacular Roman remains, for which the Society of Antiquaries of Newcastle is quite rightly known and respected.

I am sure readers will enjoy this meticulously researched and beautifully illustrated book. It should spur the enquiring mind to look differently at what might, at first sight, appear to be outwardly mundane, everyday objects and structures, so that, like the ancient surveyors, craftsmen and military quartermasters of our region, we too understand and appreciate the characteristics and qualities of the geological treasure-house that is the north of England.

The author has generously donated proceeds from the sale of the book to the Society.

David Heslop
President, Society of Antiquaries of Newcastle upon Tyne

Two heads carved from two different local sandstones, recovered from the Roman bath house excavation at Carlisle Cricket Club in 2023
(Opposite page - Mucklebank Turret, number 44B)

Preface

It is said that the Romans invaded Britain in AD 43 not only because Claudius required a triumph and it secured their Empire and influence but because they knew it was territory with abundant natural resources. There was fertile land in the south and east and to the west and north, mineral riches. Long before the conquest, in the 4th century BC, it is believed that the Greeks traded for Cornish tin and the islands became known as the Cassiterides.

This book explores what the Romans understood of the northern frontier of Hadrian's Empire, its mineral resources and the opportunities and hazards the landscape presented. The book crosses the divide between archaeology and geology.

Archaeologists and geologists share many things. They each seek to understand the past, albeit on vastly different timescales. They observe, survey and measure features and objects in the field and the laboratory. They use the same terms, like stratigraphy and provenance; and techniques, like geophysics and geochemical analysis. Their sciences are both interpretive and necessarily base their evaluations and narratives on incomplete evidence; that means their conclusions and sometimes even foundational concepts will be updated and amended. But despite all this common ground they are different. Geology is a science which aspires to resolve the puzzles rocks present and delivers a narrative based on the available evidence. Archaeology attempts to bond science and the humanities; it gathers and tests evidence but the human variable means many more narratives are created.

These two disciplines approach the study of natural resources and landscape differently. Archaeological investigation is driven by evidence of human occupation of the landscape and exploitation of its resources, whereas the geological imperative is to understand how the landscape came to be and where and what those resources are, regardless of whether they have been worked. Attempting to understand society's exploitation of geological resources on the basis of fragmentary incidences of use cannot tell the full story - absence of proof is not proof of absence. Equally, simply because geological evidence shows a resource exists, it does not follow that it has been used.

If the two approaches can be combined then our insights in this hybrid domain have a much greater prospect of fidelity.

This is not an exhaustive study of a single subject. It is a synthesis that covers many topics. It is not and does not pretend to be a specialist or academic publication, it has been designed and written for everyone curious about their landscape and its past. But because its reach is wide and crosses two domains it may provoke questions from experts within each. This book explores the Roman occupation of the northern English landscape through a geological lens, a perspective which challenges several long held archaeological views. The archaeological and historical fact base and explanations have been abstracted from a wide range of sources and the book has been reviewed by experienced archaeologists.

My scrutiny of the literature and the landscape is through the eyes of a field geologist who worked for the British Geological Survey (BGS). My job was to map the geology and identify resources and hazards. This landscape also presented the Roman army with similar opportunities and threats; building materials, mineral and water resources, effective route lines, bad ground and river crossings. Their day-to-day lives and the longer-term resilience of their infrastructure depended on their skill in evaluating and overcoming these challenges.

The book could not have been written without help and advice from many people. They are very gratefully acknowledged here and in detail at the end of the book.

Ian Jackson
7 August 2024

50 sites in 4 themes

Before the Romans

1 Yardhope Burn
2 Crag Lough
3 Moor Divock
4 Warden Hill
5 Woolaw

Reading the landscape

6 Twice Brewed
7 Sewingshields Crags
8 Peel Gap
9 Harrow's Scar
10 Heddon-on-the-Wall
11 Appletree
12 Burgh Marsh
13 Wallsend
14 Milking Gap
15 Cawfields Milecastle
16 Willowford

How do we know?

47 Catterick
48 Steel Rigg
49 Magna
50 Raeburn Flow

Natural resources

17 Housesteads
18 Fallowfield Fell
19 Combcrag Wood
20 Wetheral
21 Hardknott Fort
22 Hare Hill
23 Knag Burn
24 Newbrough
25 Cotehill
26 Scalesceugh
27 Corbridge
28 Epiacum
29 Tynehead
30 Botchergate, Carlisle
31 Burnswark
32 Brackies Burn
33 Eskdale
34 Middleton Tyas
35 Benwell
36 Vindolanda
37 Swallow Crags
38 Crosscanonby
39 Chesters
40 Hexham
41 Carlisle Cricket Club
42 Hancock Museum
43 Arbeia
44 Lanchester
45 Coventina's Well
46 Bewcastle

Bedrock geology of Northern England

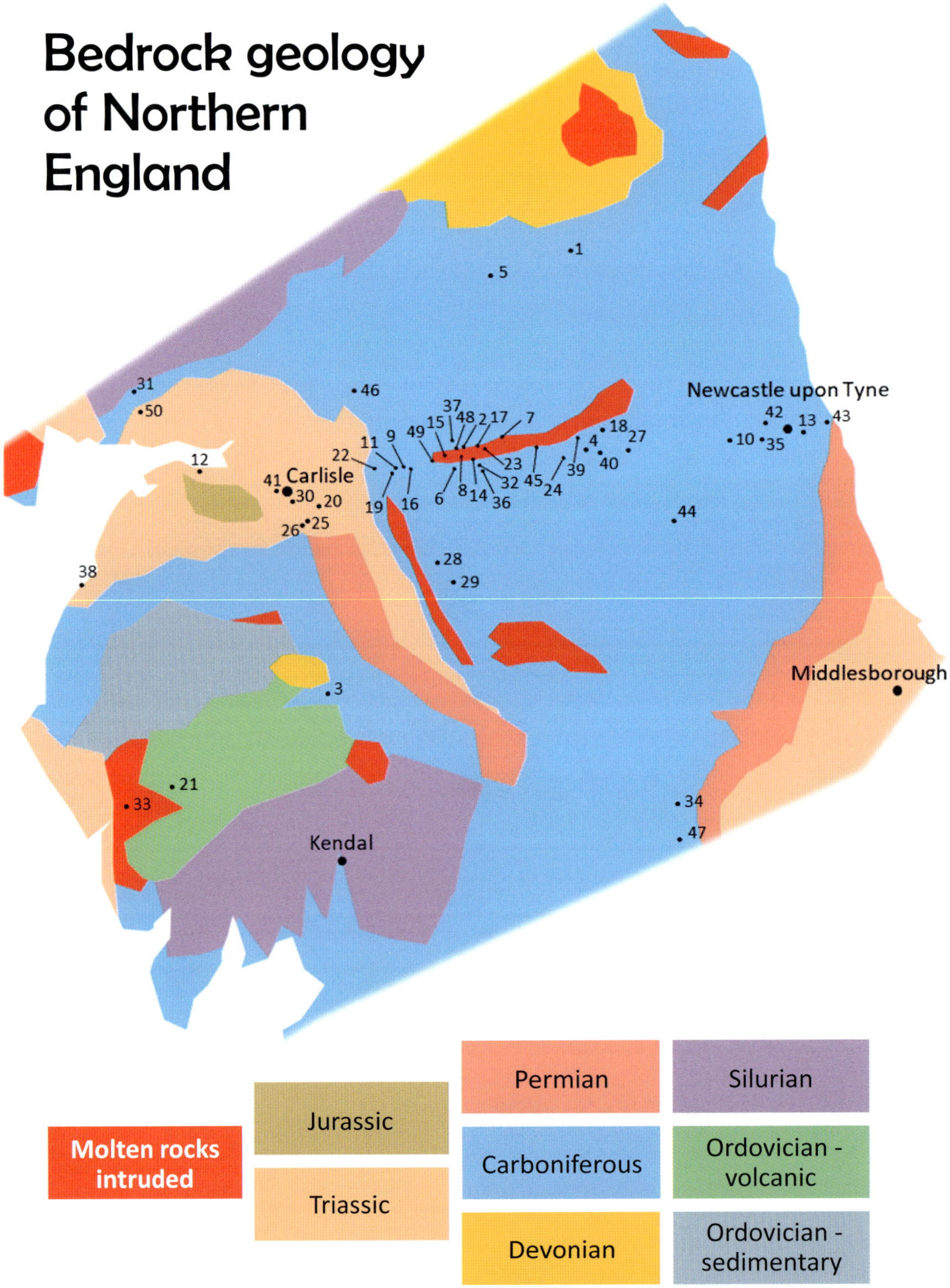

The geology of the Wall corridor

Bedrock geology - commonly buried beneath Quaternary deposits

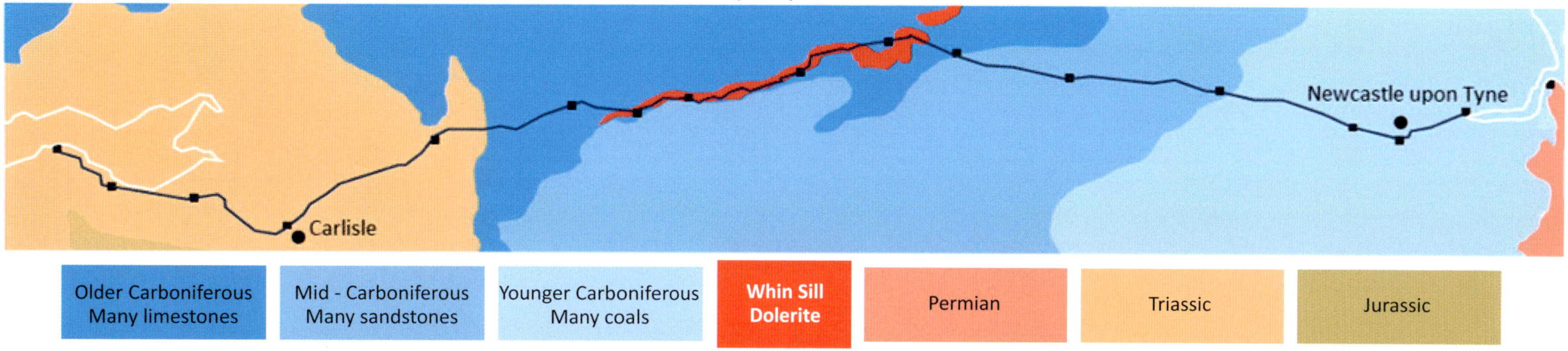

Quaternary deposits - Ice Age, river and coastal sediments and peat that cover much of the bedrock

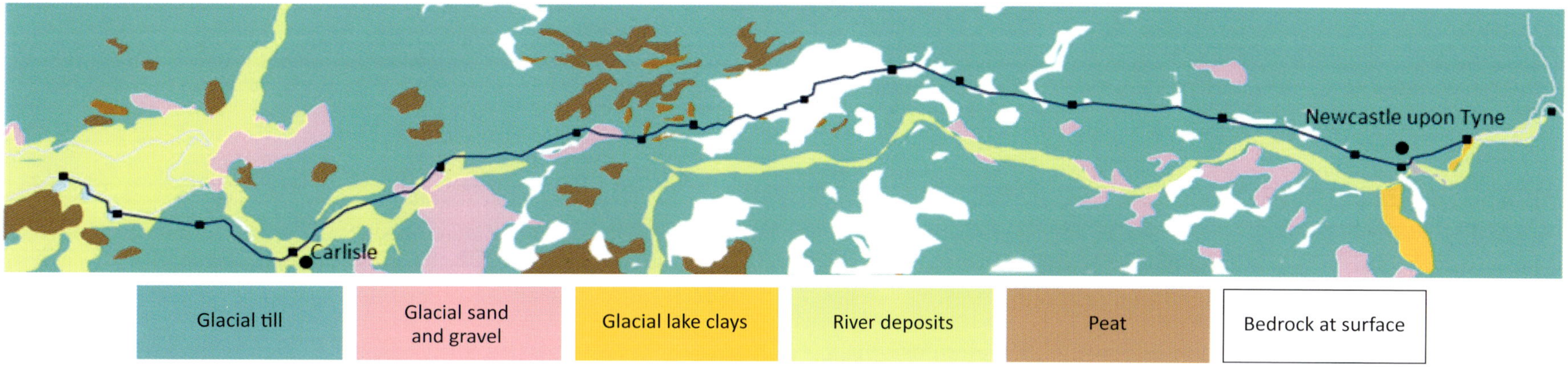

Thickness of Quaternary deposits

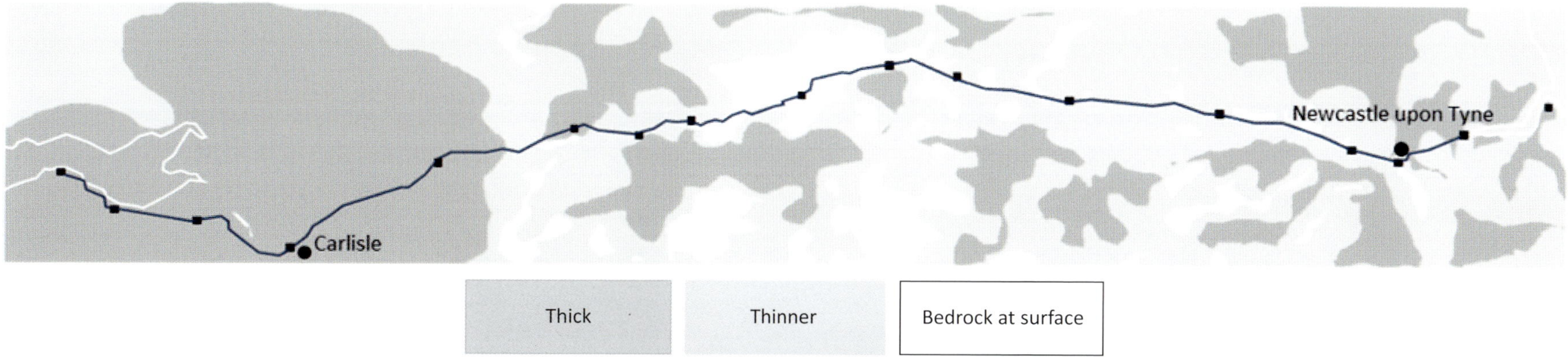

Introduction

Around 30 years after Emperor Claudius' army had arrived in Britain the Roman legions reached the north of England. Between the coastal plains of what are now Northumberland and Durham and the Vale of Eden, they encountered the remote Pennine hills and to the west, the rugged mountains of the Lake District.

These landscapes, like all others, reflect their underlying geology. The rocks beneath the northern edge of this Empire span almost 500 million years of Earth's history. What were once ancient oceans and volcanoes; swamps and coral reefs; red hot deserts and frozen wastes, are now slates, solidified lava, coal, limestone, sandstone and clay. The unrivalled diversity of the shape and composition of these northern rocks provided opportunities and riches, but they also created terrain that would challenge the Roman occupation.

The legions did not conquer an uninhabited landscape. Stone, then Bronze, then Iron age peoples had been here for at least 8000 years. They had cleared much of the primary forest, planted crops and exploited the geological opportunities too: stone and metal ores for tools, ornaments, buildings and monuments; hilltops for defence and burial; springs and water for life and faith.

This book is no more than an introduction to the way geology shaped the occupation of the most northerly part of Hadrian's Empire. Through stories about 50 different places the book explores the landscape the Romans first encountered. How did geology influence their decisions about infrastructure and how did they exploit and trade the resources it offered? It also introduces the relevance of geology to the dating, provenance and preservation of their remains and environment. These stories have been collected into four themes: Before the Romans; Reading the landscape; Natural resources; and How do we know?

Greenlee Burn, north of The Wall

The components of Hadrian's Wall – a very brief explanation

It hasn't always been Hadrian's Wall, until 1840 it was assumed to have been built by the Emperor Septimius Severus. It hasn't always been stone either, the western 50 kilometres was initially built of earth and is called the Turf Wall. Hadrian's frontier is much more than just a wall, to its south are a ditch and ramparts called the Vallum and sited along its length 16 forts, 80 milecastles and 160 turrets. A road, the Military Way, serviced all of this infrastructure.

Geological time - in millions of years

0	
Quaternary	Multiple ice ages and warm periods and the beginning of humankind
2.6	
	No Cretaceous or later sedimentary rocks remaining in northern England
145	
Jurassic	Seas teeming with life but only a small outlier remains
201	
Triassic	Flood prone rivers and transient lakes in a desert environment
252	
Permian	Migrating sand dunes then inundation by coral seas
299	
Carboniferous	Cycles of rivers running through tropical swamps into coral seas
359	
Devonian	Erosion of a large mountain chain and deposition of gravel and sand
419	
Silurian	An ocean between two colliding continents filling with sediment
444	
Ordovician - volcanic	Violent volcanic eruptions producing huge amounts of ash and lava
450	
Ordovician - sedimentary	Primitive life in a deep ocean that was filling with mud and sand
485	

Molten rocks intruded

Archaeological time - years AD and BC

Roman army leaves northern England	~AD 407
Garrisons lose contact with Rome	~AD 300-400
Severus rebuilds Hadrian's Wall	~AD 200
Antonine Wall built	~AD 142
Hadrian dies, Antoninus Pius emperor	AD 138
Hadrian's Wall built	~AD 122
Hadrian made emperor	AD 117
Stanegate road built	~AD 80
Roman legions reach Carlisle	~AD 72
Defeat of Boudicca and southern tribes	AD 60
Claudius invades Britain	AD 43
Julius Caesar first lands in Britain	55 BC
Roman era begins	785 BC
Iron Age	800-1 BC
Bronze Age	2,200-800 BC
Neolithic	4,300-2,200 BC
Mesolithic	9,000-4,300 BC
Palaeolithic	800,000-9,000 BC
The first humans	~2,500,000 BC

1

Yardhope Burn

What did the northern landscape look like when humans arrived?

Let's wind back 20,000 years. Northern Britain was, not for the first time, covered by an ice sheet several hundred metres thick. Approximately 15,000 years ago that ice sheet started to melt away leaving an ice-sculpted landscape and a veneer of stony clay, silt, sand and gravel. The white waste and bare ground were quickly replaced by tundra with mosses and lichens and as it got warmer, grassland prairies and woodlands, but the warmth didn't last. Around 13,000 years ago our climate cooled again and ice caps formed on the higher ground in the Lake District and Scotland. This mini-ice age and associated tundra conditions affected northern Britain for about 1500 years. The ice disappeared as quickly as it came and plants and trees – and humans – returned.

People, including Neanderthals, had been here before, possibly from about 800,000 years ago. Every time one of the multiple ice ages came to an end and the climate ameliorated, early humans would return to forage and hunt. This was the Palaeolithic (or 'old' Stone Age) period, so-called because of their stone tools. The Mesolithic ('middle' Stone Age Period) began approximately 11,000 years ago when the freezing conditions ended. These hunter-gatherer people initially moved into a landscape of mosses, grassland, scrub and reindeer that, as the climate warmed up, was again progressively replaced by birch, hazel and pine woodland, followed by oak, elm and alder. The widely accepted story used to be that by 7000 years ago the lowlands and valleys were covered by broadleaved woodland and inhabited by red and roe deer, aurochs, horse, wild pig, bear and lynx. In the 21st century the idea of closed canopy woodlands has been challenged – did those early herbivores keep the trees at bay and was the landscape actually one of wide areas of grassland interspersed with woodland?

Today neither of those habitats exist, but one of the places you can get a feel of similar ancient woodland and enjoy its rare tranquillity is in the valley of Yardhope Burn and its tributaries, west of Holystone. The last ice sheet left stony clays covering much of the land in and around this valley but it also eroded and sculpted the bedrock – 340-million-year-old Carboniferous Fell Sandstone – and made it more prominent. If you visit Yardhope you will glimpse this sandstone outcropping on the valley side. In a niche in the exposed bedrock there is a Roman shrine to a local god, Cocidius. See if you can find it.

The shrine to Cocidius south of Yardhope Burn

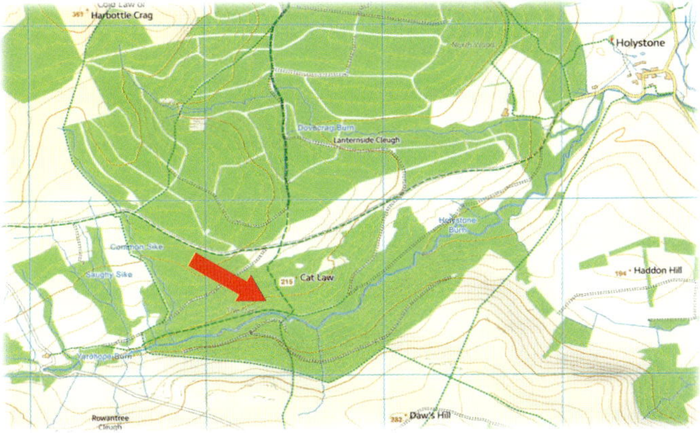

Park at the Forestry England carpark 1.5 kilometres west of Holystone and walk west on the footpath towards the burn [NT933014]

Ancient woodland in the Yardhope Burn valley

Crag Lough

Crag Lough is the best known of Northumberland's 16 loughs. It nestles beneath one of the highest and most rugged sections of the Whin Sill.

The Lough (lake) sits in a long shallow depression that was scooped out by ice sheets flowing west to east across the country. It used to be bigger and drained west through the Peel Gap channel. Over thousands of years sediment gradually accumulated on the lake floor. That sediment included mud, silt and wood brought in by water but also pollen and other organic material falling from the air. To learn about past environments scientists examine such sediments and pollen in microscopic detail and determine their age using radiocarbon dating. They also take into account the shape of the land and the height of the sea, the remains of humans and their artefacts and use all these inputs to paint a picture of how our landscape has changed through time. The beginning of human-induced change from natural woodland to a more diverse, patchwork landscape is believed to have started slowly around 6000 years ago and accelerated. Instead of relying on hunting and gathering, Neolithic (New' Stone Age) people had a different lifestyle – the stalking of prey was now coupled with subsistence farming. Progressively larger areas of woodland were cleared for pasture and crops.

Until 4500 years ago the tools were of stone, but then metal tools begin to appear. Initially they were of copper, but that was soon replaced by bronze. By 800 BC our ancestors had learned how to smelt and work iron. People grouped together in regional tribes, more wood was needed for fuel, more trees were felled and more food was grown and goods traded. 500 years before the Romans arrived there was a large-scale opening up of the landscape. It was this varied mosaic of pastoral, arable and woodland that the Romans would inherit.

Just as today the environment in Roman times was diverse. It was influenced by the climate, by the elevation of the land and the rocks beneath it. Low land along the valleys and coasts was occupied by subsistence farming interspersed with woodland. On the thin soils of the uplands scrub and moorland dominated. The highest land, like the Lake District, has been assumed to have been above the treeline, but current rewilding ventures are hinting that, unless they are covered by blanket bog, trees are able to colonise even those. The Hadrian's Wall area has had more prehistoric pollen data analysed than perhaps any other part of Britain. These data are not infallible – results from one bog cannot be extrapolated to the wider landscape – but sediments and pollen from places like Crag Lough still give us the best picture of how our climate and vegetation has changed since the last ice melted away and humans evolved.

Crag Lough from Hotbank Crags

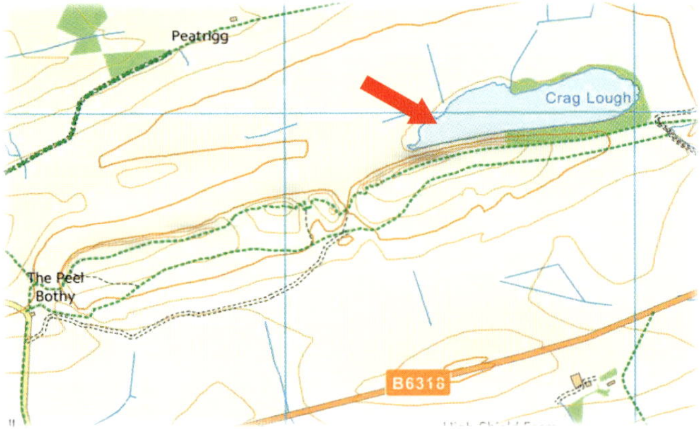

Walk 2 kilometres east along the Wall Trail from Steel Rigg car park [NY765679]

Crag Lough from the west

Moor Divock

Near Askham, on the limestone rim that encircles the Lake District, there's a place littered with stone monuments thought to be from the Neolithic and Bronze ages.

Stone circles, standing stones, burial cairns and chambers ('tumuli' and 'cists'), earth banks, and ring and field-clearance cairns are spread across a plateau of rough common land underlain by a rock called the Shap Village Limestone. This was deposited in tropical Carboniferous seas 345 million years ago. Despite the veneer of till (stony clay deposited by the ice sheets 20,000 years ago), the limestone beneath is still evident: the sink, or shake holes (karst features) give it away. While geology shaped the terrain, whether the bedrock had anything to do with our ancestors choosing this location is unknown, but limestone does influence the vegetation and the ice sheets certainly delivered the erratics that form the monuments ....and could the sink holes play any role in ritual?

There are a lot of speculative theories about the man-made monuments but exactly why people built them and buried their dead here is not well understood. Many of the stone monuments have names. The Cop (or Kop) Stone, an erratic of Ordovician volcanic rock, stands alone in the southeast. To its north is a ten stone circle around a turf mound known prosaically as Moor Divock 4. Excavation of the mound of this ring cairn in 1866 revealed an urn and human bones. The Cockpit is more evocatively named. It is a circle of at least 20 large and many more small stones of diverse geological origin. 700 metres to its northeast is White Raise, a 'star' cairn with a coffin-shaped stone-lined burial cist you can peer into. Dotted around are many other clearance cairns, rows of stones and solitary ones.

Transitions between human eras are, to use a geological term, diachronous, they happened at different times across Britain and the world. Changes in human habitation in Cumbria would have been equally uneven and coastal lowlands with their food resources and ease of travel would be occupied well before the mountainous and wooded interiors. The Neolithic people who erected the first stones probably arrived in this area about 6000 years ago. The transition to the Bronze Age societies who are thought to have added monuments was around 4000 years ago. There is another, later, twist in the Moor Divock tale. High Street, the well-known Roman Road, descends from the fells to the south and west. It passes close by the Cockpit stone circle. However, despite its direct line and near legendary status, there are those who, noting its atypical form and the fact that it does not join Roman sites, ask: could it just be a much later packhorse track?

The Cop (Kop) Stone on Moor Divock

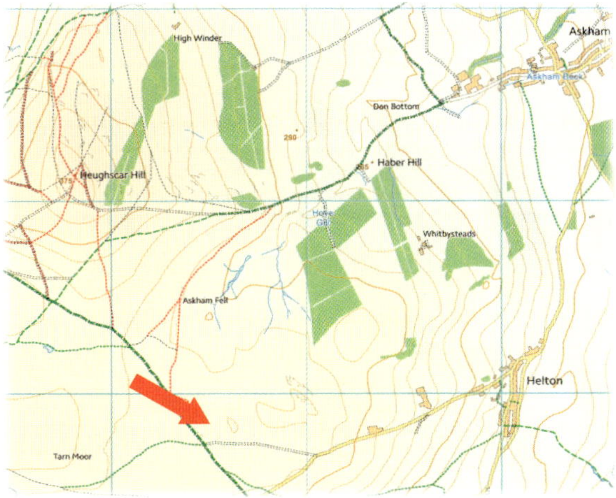

Park beside the minor, no-through road between Helton and Scalegate Farm and walk north [NY495218]

Moor Divock 4 stone circle

4

Warden Hill

The Romans did not encounter a wilderness when they advanced into northern Britain; this was already a managed landscape. Iron Age settlements and hill forts were everywhere. Warden Hill, northwest of Hexham, is one of those forts.

There is another fort just to the west at Barcombe Hill and further northeast and northwest spectacular large forts such as Yeavering Bell and Burnswark. More subtle but much more numerous are the traces of Iron Age farms and settlements being identified across the region by aerial photography and new techniques like LiDAR (Laser imaging Detection And Ranging: measuring ground height by laser from the air) and high-resolution geophysics.

The fort at Warden sits on the flattish top of a hill with commanding views overlooking the valleys and confluence of the North and South Tyne. The remains of the multiple stone and earth ramparts are now spread and subdued but still easily seen, as is the western gateway. There has been no modern excavation but the fort may enclose circular houses and there are also signs of later structures, a possible Romano-British settlement and, to the east, traces of a later settlement. Britain during the Iron Age was divided into a collection of amorphous tribal regions. Hadrian's Wall apparently cut through the land of the Brigantes and encroached on that of the Votadini in the northeast. To the west their neighbours were the Selgovae. Many chiefs collaborated with the Romans (some becoming "client kingdoms") but individual tribes were not homogeneous units and relationships were not always friendly. These Iron Age peoples had detailed local understanding of their territories – the geography, topography and natural resources and it is likely that the Roman army benefitted enormously from their knowledge of the landscape and its resources.

Warden Hill is made of flat-lying Carboniferous rocks around 325 million years old. They have been eroded and sculpted by the action of an ice sheet just 20,000 years ago. At the top is sandstone and below that mudstone, then more sandstone and mudstone and a limestone. Descend the hill along the western path from its summit; its steep and less steep sections reflect those alternating harder and softer rocks, each older than the last. With each layer you are going back tens of thousands of years in time.

The west gateway through the ramparts of Warden Hill fort

Park in Fourstones. It's a 3 kilometre walk there and back [NY904678]

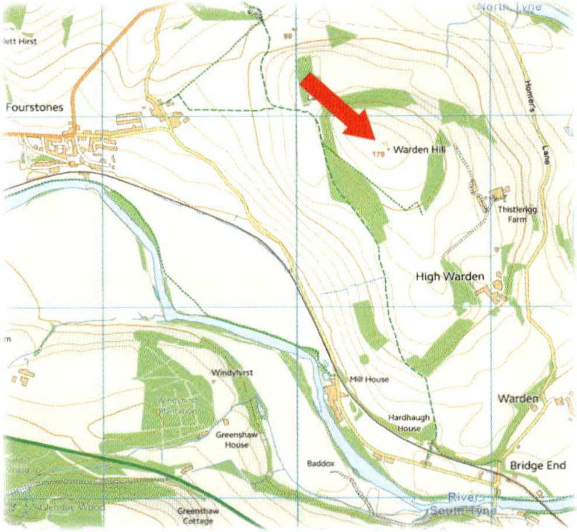

Warden Hill fort looking southeast.
Image kindly taken by David Barnes.

Woolaw

Most of the Iron Age settlements the Romans encountered are largely lost or invisible to us now. At Woolaw, near Rochester, the remains of a typical farmstead survive.

There are Iron Age sites across the uplands of northern England: defended settlements and homesteads. There were many farmsteads in the lowlands too, but multiple changes to the landscape made by successive societies over the last 2000 years mean that accessible and well-preserved examples of the homes of these indigenous people are rare. On a knoll of exposed Carboniferous sandstone south of the River Rede are the remains of four, stone, roundhouses within a rectangular earth and stone embankment, still standing almost a metre high above the ground. The houses are between four and seven metres across and each pair has a yard in front.

This site is not an isolated piece of archaeology. A few hundred metres to the northwest and southeast are three more Iron Age/Romano-British farmsteads, with areas of 'cord rig' (parallel raised ridges of soil for cultivation). These settlements are described as Romano-British because many evolved long before the Romans arrived and were used all the way through the period of occupation. Recent research has, however, concluded that many Iron Age settlements immediately north of the Wall in south Northumberland were abandoned soon after it was built. Whether they were cleared to allow Wall construction, or the barrier meant their way of life was no longer possible, is unknown.

Nearby Rochester has a great deal of evidence of Roman occupation too. The village lies in the valley of the River Rede, a natural line for their road, Dere Street, leading up to the crossing of higher ground of the Cheviot mountains formed by older (and harder) Devonian volcanic rocks. Temporary camps to Rochester's northwest testify to recognition of the strategic importance of the route it protects. There is a fort (Bremenium) just north of the village which was established around AD 80 as part of Agricola's advance into southern Scotland. It was replaced by a more substantial earth, turf and timber fort about AD 140 following the decision by Antoninus Pius to extend the Empire to the Clyde-Forth isthmus and possibly abandoned, to be later rebuilt by Septimius Severus in the early years of the 3rd century. Good quality building stone, Carboniferous sandstone, is abundant and close to the surface nearby, but linking local small quarries to Roman extraction would be tentative at best. Many stones from the fort are incorporated into 19th century buildings nearby, including guttering and 'ballistae' – stone projectiles!

Rochester Old School House porch gable with reused stone from Bremenium Roman Fort

Park at Bellshiel Bridge. The settlements are on Access land 1 kilometre south [NY815984]

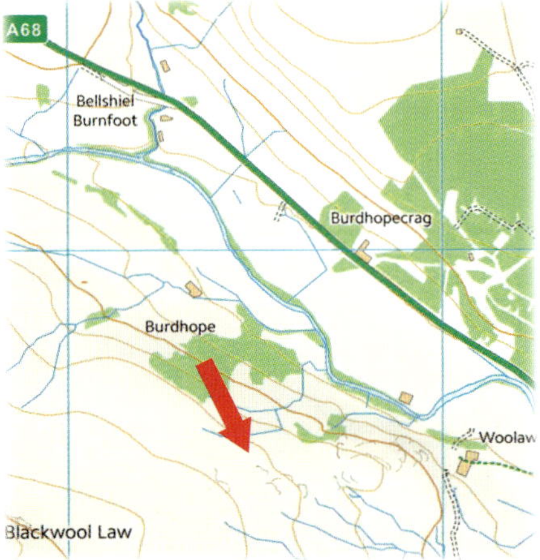

Woolaw Iron Age (and Romano-British)
round houses and settlement looking east

Twice Brewed

In parts of Scotland, Wales and the north of England low banks of earth in the shape of rectangles can often be spotted in the fields. South of the central sector of Hadrian's Wall they seem to be everywhere.

Roman camps don't look much, after almost 2000 years most of their earthworks have spread and are now just low linear mounds. The one immediately south of the Twice Brewed Inn is just like that and so is best seen when the sun is low and shadows are long. Then you can pick out the ramparts and the gateways, often guarded by additional defences. The influence of the underlying geology on temporary camps like these is not that significant, although you don't see many where bedrock is near the surface, like the steep southern dip slopes of the Whin Sill dolerite and sandstone scarps. Their builders seemed to prefer to locate them on softer ground – glacial stony clay (till or boulder clay). They sensibly avoided what are now wet peat bogs (which may have been lakes then) too. Many camps are close to roads, like the Stanegate, logical as soldiers prefer to march on a firm base.

The mounds seem to have been dug or scraped up from the exterior leaving a complementary ditch and together these may have produced a rampart originally around 1.5 metres high. There is little detail on their construction as very few modern excavations have been done. Roman military manuals advised they should be built with their front facing the enemy and their rear close to the crestline to avoid dead ground. Many camps are on slight slopes – not ideal as those of us who have slept in tents will testify – but a strategic-defendable site not comfort was the priority. That over-riding factor, a safe position and perimeter, also means that the size of the camps does not necessarily relate to the numbers of soldiers they contained; a topic of archaeological debate for a long time.

The camps vary considerably in size. 'Seatsides 1' Camp less than 400 metres south of the Twice Brewed Camp is over 6.7 hectares in area (252 x 282 metres). It and its near neighbour 'Seatsides 2', could have held many thousands of men. In contrast 400 metres southeast of 'Seatsides 1', beside the road to Bardon Mill, is a camp which measures barely 25 x 25 metres. The size of these smallest camps may well indicate their purpose: troop training – army habits do not change! But the function and dates of the larger camps have been less easy to establish and much conjecture surrounds them. Some will undoubtedly have been for construction of the Wall and its infrastructure but arguably the majority were built by army units on manoeuvres.

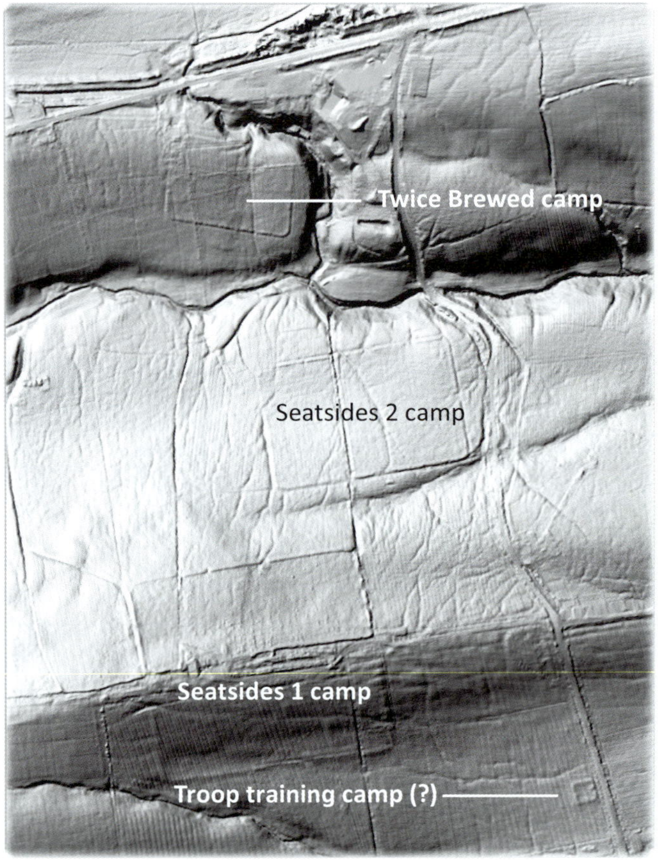

Twice Brewed and nearby camps.
Digital surface model at >1 metre vertical accuracy.
Environment Agency data processed by GIS software.

The camp is in the field south of the pub beer garden [NY751668]

The Roman camp south of the Twice Brewed Inn

Sewingshields Crags

The view is dramatic and stretches away to the west to the high point of the Wall on Winshield Crags and beyond. But the feature that dominates the landscape is not the Wall, it is the escarpment of the rock that underpins Hadrian's monument: the Whin Sill.

This 30-metre-thick slab of molten magma was injected at around 1200 °C into older Carboniferous sedimentary rocks 295 million years ago. The escarpment you see is just the tip of the Whin Sill. It extends for over 4500 square kilometres beneath the Pennines of Cumbria, Durham and beyond. It is England's biggest igneous intrusion – a rock called dolerite, made up of mineral crystals just visible to the naked eye. Zoom out to a rock face and it shows the characteristic vertical columns caused when the magma cooled and contracted. At landscape scale, its hardness and resistance to erosion creates the prominent escarpment. 150 years ago, an Alnwick draper, postmaster and polymath, George Tate, aided by scientists of the Geological Survey, proved definitively that it was not volcanic lava extruded over the surface of the ground but an injection of molten rock deep underground and exposed now because of millions of years of erosion.

If you'd like to see the evidence there is limestone altered by its heat (metamorphosed) above the dolerite at Sewingshields and further west, barely 20 metres north of Hotbank Farm, the footpath crosses the mudstone directly beneath the Sill which has been baked to a slatey rock.

Current thinking on Hadrian's decision where to site his Wall is that he had the choice of three locations to separate his Empire from the barbarians: the Clyde-Forth, the Solway-Tyne or the Mersey-Humber. If he had designs on the mineral wealth of the north of England, he had just two. He chose the Solway-Tyne. Was that because of the pre-existence of the Stanegate road or the strategic value of another bit of geology, the Whin Sill? According to recent research it was the 28-kilometre section of the Sill's north-facing crags that dictated the route of the Wall corridor across the isthmus. The Wall's east and west limbs to the respective coasts were subservient. They merely stay on higher, drier ground and avoid streams and rivers as much as possible.

Take a walk along the Wall from Sewingshields to Hotbank Farm and then cross the frontier and follow the footpath north. Turn west after 300 metres and head to Steel Rigg. Now turn your gaze to the south. On a clear day the views of the Whin Sill will amaze you. But even on a dull day it is magnificent and moody. It is the rocks that are the foundation of these views. They define Wall Country and they defined its Wall. We humans are but latecomers.

Sewingshields Crags and the characteristic Whin Sill escarpment with its dolerite columns

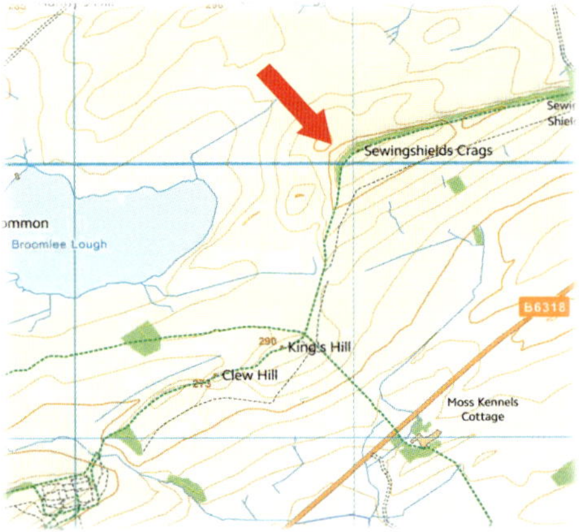

Walk 2 kilometres east along Hadrian's Wall Trail from Housesteads [NY800700]

The Whin Sill escarpment from Sewingshields Crags, to Housesteads, to Winshield Crags and beyond

Peel Gap

Walking along the middle bit of Hadrian's Wall with all its ups and downs can be hard on the leg muscles. Ever wondered why these undulations are there?

From Sewingshields to Walltown the Wall follows the crest of the Whin Sill, a hard igneous rock called dolerite that has been shaped by ice to produce its formidable escarpment. But the escarpment isn't level, it undulates repeatedly and sometimes steeply. Geologists believe that many of these short little valleys, cutting generally north-south, are places where the Whin Sill has been weakened, sometimes by faults – dislocations in the rock caused by Earth movements hundreds of millions of years ago. Faults are rarely single planes, they are broken and shattered zones which are more easily eroded.

Fast forward to the Ice Age which began less than 2.6 million years ago. On more than one occasion several hundred metres of ice covered all of Scotland and northern England. When these ice sheets began to melt the water beneath them tried to escape southwards to the lower ground – the valley of the Tyne. The escarpment of the Whin Sill was initially a barrier to the meltwater but the weaker zones in the rock provided opportunities for water, under considerable pressure and with a lot of debris, to break through. That is how we think the gaps formed – as meltwater channels. Some of them may have been subglacial as their profiles are 'humped'; others may have drained glacial lakes of ponded up meltwater.

Like Peel Gap most of these channels – gaps, nicks and slacks – have names. One, Sycamore Gap, is world famous (sadly now because of an act of vandalism). Some like Rapishaw Gap, Thorny Doors and Lodham Slack are less well known. Others like Busy Gap – said to be a gathering place for Reivers and rogues – have romantic stories, but each channel has its own character. By the time you have hiked up and down them (others are Castle Nick, Caw Gap, Green Slack, Milking Gap and Hole Gap) their cumulative impact on your leg muscles will be apparent. Take a walk east from Steel Rigg car park and you soon descend into Peel Gap. Its origin is a boggy valley west of Crag Lough and then it turns south to run through the Wall. In the gap there are the foundations of an extra Wall tower between turrets 39A and 39B. It is simply butted onto the Wall which suggests it was built after the Wall, perhaps to defend this unsighted low point, or maybe to provide soldiers with shelter and access to the Wall. Tackle the Wall's geological ups and downs and you will soon appreciate why the Roman army built the Military Way across less daunting terrain to the south.

Castle Nick and Sycamore Gap (in 2023)

Walk 150 metres east from Steel Rigg car park [NY753675]

Peel Gap looking south

Harrow's Scar

The position and initial structure of the Wall and the Vallum were defined by geology – it is the dominant influence on both the landscape, foundation conditions and building materials.

If you were to roughly divide up the Wall corridor in terms of 'rocks' there are 3 main chunks and the surface landscape reflects these. There is an obvious middle third, roughly 28 kilometres from Greenhead to Chollerford, defined by the hard dolerite of the Whin Sill. The eastern third to Wallsend is Carboniferous sedimentary rocks covered by a patchy veneer of glacial stony clays. The western chunk is dominated by thick glacial and more recent river and coastal sediments which, except in river valleys, hide the bedrock. It is conventional to stop and start the central and western sectors of the Wall at the 'Red Rock Fault', a line going approximately north-south through Lanercost which divides western red from the eastern grey-brown bedrock. This is not a fault at all, it is an unconformity (a fault is a tectonic fracture of the rocks, an unconformity is a gap in the rock record). This unconformity is where water depositing Permian and Triassic sediments encroached progressively eastwards on the eroded surface of Carboniferous rocks around 270 million years ago.

The reason archaeologists chose to draw the dividing line here was that this is where the bedrock (the source of the building stone and mortar) changes. However, because that change is buried deep, this is not where geology influenced the initial alignment and construction of the Wall in practice. That is around 10 kilometres to the east, at the River Irthing, where the Quaternary (or Superficial – modern geological maps no longer use the word 'Drift') deposits thicken considerably. It could be even further east, at the Tipalt Burn, where the valley of a much larger pre-Ice Age river, the proto-Tyne, is plugged by tens of metres of glacial sediments. From here to Bowness-on-Solway it is the unconsolidated Quaternary deposits, not bedrock, that are the dominant influence on the shape and composition of the terrain – and on Roman plans.

West of this point it would no longer have been possible for Roman builders to simply open multiple small, near-surface, sandstone and limestone quarries close to the Wall. The building materials that are now readily accessible are Quaternary deposits – clay, stones, soil ......and turf. Walk 400 metres east from Birdoswald Fort and a few metres more beyond Harrow's Scar Milecastle and you encounter the 20-metre-high river cliff of the River Irthing. From top to bottom this cliff is made of unconsolidated stony glacial clay, gravel and sand, Carboniferous bedrock is buried deep beneath it. West of the milecastle the original Turf Wall starts its march to the Solway – the change in building material seems an unlikely coincidence.

East gate of Birdoswald Fort looking towards Harrow's Scar Milecastle (49). A Roman road ran on the line of the Turf Wall up to the gate.

Use Birdoswald fort car park and walk east along Hadrian's Wall Trail for 500 metres [NY621664]

Thick glacial gravel, sand and clays overlying Carboniferous bedrock (bottom right) just upstream of Harrow's Scar

Heddon-on-the-Wall

From Newcastle to the crossing of the River North Tyne the Wall was built in long straight lengths in stone, adjusting its direction on hilltops.

What, if any, influence did geology have in this eastern sector of the frontier? This terrain is very different to that of the central, Whin Sill, sector. The bedrock here – layers of sandstone, mudstone and limestone – tilt more to the east than the south and the resistant Whin Sill swings away northeastwards. Just west of Heddon, the Wall crosses the dividing line between the older Carboniferous rocks and the younger rocks of the southeast Northumberland Coalfield. This division is based on fossils and a progressive change from open sea to coastal swampland – which produced abundant coal seams. This change cannot really be recognised in the rocks at surface but from here on eastwards the evidence of the last three centuries of coal mining gives it away, even today.

But the 320-million-year-old bedrock is not the only influence on the landscape here; a much more recent geological event shaped this land too. Only 20,000 years ago a 700-metre-thick ice sheet flowed eastward sculpting the rocks beneath it. The glacial debris it left behind smoothed out the terrain even further. The imprint of the last glaciation here is west-east ridges – resistant low sandstone hills peeping through a thin covering of stony clay plastered unevenly across an undulating landscape. The line the Roman Wall planners chose used those high, dry ridges, all the while trying to maintain a view into the valley of the Tyne to the south. Benwell, Heddon, Eppies Hill and Harlow Hill are four of those higher places, each one underlain by sandstones, two of which have been quarried (at least in post-Roman times).

In Heddon village sandstone is exposed beside the line of the frontier and a little further east over 200 metres of the Wall and its ditch are visible. This fragment is one of the best-preserved sections of the originally specified 'Broad Wall'. It had a theoretical width of ten Roman feet (~3 metres). Soon after its building in AD 122/123 the specification changed and the thickness of the Wall was reduced to 'narrow gauge' (~2.2 metres). A thinner wall would speed up construction and was synchronous with a decision to place 12 forts and more than 6000 men along the frontier. 100 years ago a section of the Wall rebuilt by Severus at Great Hill in Heddon was excavated so a new road could be built. The core of the foundation level between large, quarter ton, sandstone blocks was cemented stone rubble, set so hard it had to be blasted out. Another Roman rock recipe – concrete.

Looking east to Eppies Hill along the Military Road (B6318), which is built on top of the Wall

The outcrop is on the Towne Gate in Heddon along the line of the Wall [NZ136670]

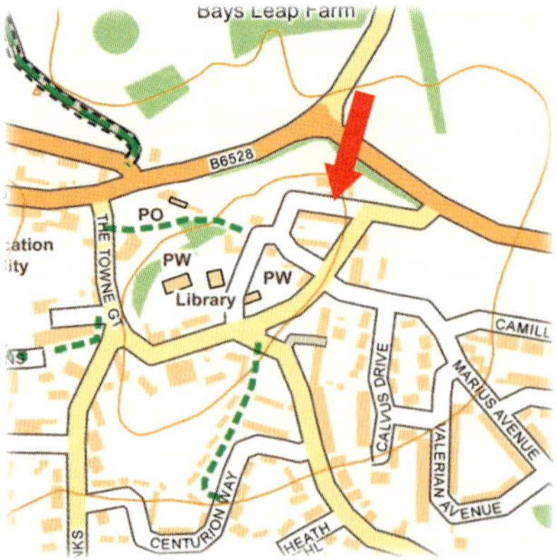

Stone wall blocks (some of which are Roman) on top of natural Carboniferous sandstone with sloping 'cross bedding' - old dunes deposited in a river

Appletree

It's called the Turf Wall. But the western third of Hadrian's Wall is more of an earth wall.

There is turf on the outside but the core is more soil, subsoil and clay. It is just one of a series of defensive 'Earthworks' so it qualifies for inclusion in a book about Romans and geology. Appletree is one of few places where you can easily see all the components of Hadrian's linear frontier, but they can get confusing.

The Stone Wall ditch and bank are just north of the modern road and the Stone Wall is beneath it. The Turf Wall defences begin 100 metres to the south. First is a low mound of material that slopes gently north, then the Turf Wall ditch and the Turf Wall. Less than 20 metres south of the Turf Wall itself are the three mounds and ditch of the Vallum. The low mound in front of the Turf Wall ditch is 16.5 metres wide but only 0.5 metres high and is made of clay with some stones. It lies directly on clay as the turf and subsoil which were once above had been stripped for building the Turf Wall. The ditch, almost 11 metres wide and three metres deep, is cut in glacial stony clay (geologists call this till or diamicton), the excavated material will have been used for both the mound and the Turf Wall. The Turf Wall was six metres wide and four metres high but was levelled by the Romans when they built the Stone Wall, so a broad bank only 0.5 metres high remains now. It had a near-vertical north face and a sloping south one. There may have been a walkway along the top and a woven wooden hurdle parapet too. Excavations here in 1999 and elsewhere proved the Turf Wall to be a mix of turf, soil, subsoil and clay and sometimes wood and stones. At its base they found a black humic layer around four centimetres thick – the old grass surface; the turf had not been stripped.

The Turf Wall provokes as much debate as the Stone Wall. Did the proportions of its building materials change along its length too? The question of why the western sector was first built of turf exercises a lot of minds. Was it an expedient solution to the threat of troublesome tribes, or a logical use of the available raw materials, or both? On Roman frontiers across the Empire turf and wood were often used. How long did the Turf Wall survive before it was replaced with the Stone Wall? Evidence from Birdoswald Fort suggests that here it may have been within Hadrian's reign, further west probably later. Walk towards the Appletree section from Birdoswald, you get a fantastic view of these Earthworks and their horizontal scale. To appreciate their vertical dimension, there are full-scale reconstructions at Tullie House museum in Carlisle and Vindolanda.

Turf Wall reconstruction Tullie House.
Image taken by kind permission
Tullie House Museum & Art Gallery Trust.

Walk west for two kilometres along the Hadrian's Wall Trail from Birdoswald [NY600657]

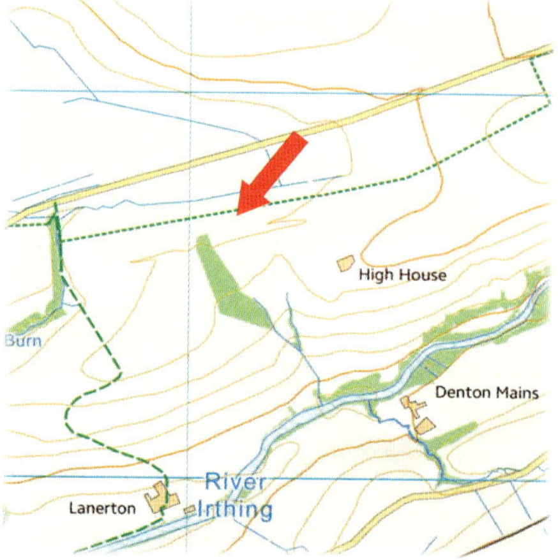

The Earthworks at Appletree

Burgh Marsh

Hadrian's frontier crossed the River Eden at Carlisle and 10 kilometres beyond met the Solway estuary. How did his legions deal with this dynamic geological environment?

West of Birdoswald Fort the Wall follows subdued ridges and descends slowly into the Solway basin. The bedrock, Carboniferous in age to Lanercost, then Permian and Triassic, is mostly deeply buried by deposits of the last glaciation. These stony clays, silts, sands and gravels were the product of the advance and subsequent melting of an ice sheet over 700 metres thick. The history of the last glaciation in the Solway basin is complicated and the evidence from glacial erratics, geological deposits and topographic indicators of ice flow is conflicting. From 28,000 years ago there were multiple ice advances and retreats until 15,000 years ago when the area finally became ice-free.

At that time sea level was depressed by perhaps 20 metres below today because of the amount of the world's water locked up in the ice. The weight of ice had also depressed the land and so when the climate began to warm the release of ice pressure allowed the land to 'rebound' (isostatic change). The melting of the ice discharged huge amounts of water into the oceans, raising their level (eustatic change). These opposing changes didn't happen equally everywhere, nor at the same time. Sea level in the Solway initially rose to around two metres above today's level and then fell to around 0.5-1 metres above it in Roman times.

How did these changes affect the Roman frontier? After bridging the Eden near its confluence with the River Caldew in Carlisle, the Wall follows the Eden's southern bank, staying on firmer glacial stony clays and avoiding the lower, riskier river alluvium. It then heads directly west and between Beaumont and Dykesfield continues on glacial deposits, moulded into kilometre-long parallel ribs by the movement of the ice. From Dykesfield to Drumburgh there is no Wall or Vallum. The terrain here is flat saltmarsh, only a few metres above sea level and regularly flooded by high tides. That would have happened more often when the sea level was higher 1900 years ago. The geology here is soft, wet silt and clay – raised tidal flats often called 'warp'. These sediments are still being deposited (and eroded) by the sea. Geology and sea level explain the absence of frontier installations here. When the Wall reappears at Drumburgh it and the Vallum take full advantage of drumlin-like hills, appearing today as islands of glacial material surrounded by tidal flats. The defences hop northwest, from Glasson to Port Carlisle and finally to Bowness-on-Solway, ensuring oversight of the most westward fording point of the Solway Firth.

Soft tidal deposits being eroded at the edge of Burgh Marsh

The marsh is accessible between Burgh-by-Sands and Drumburgh [NY289600]

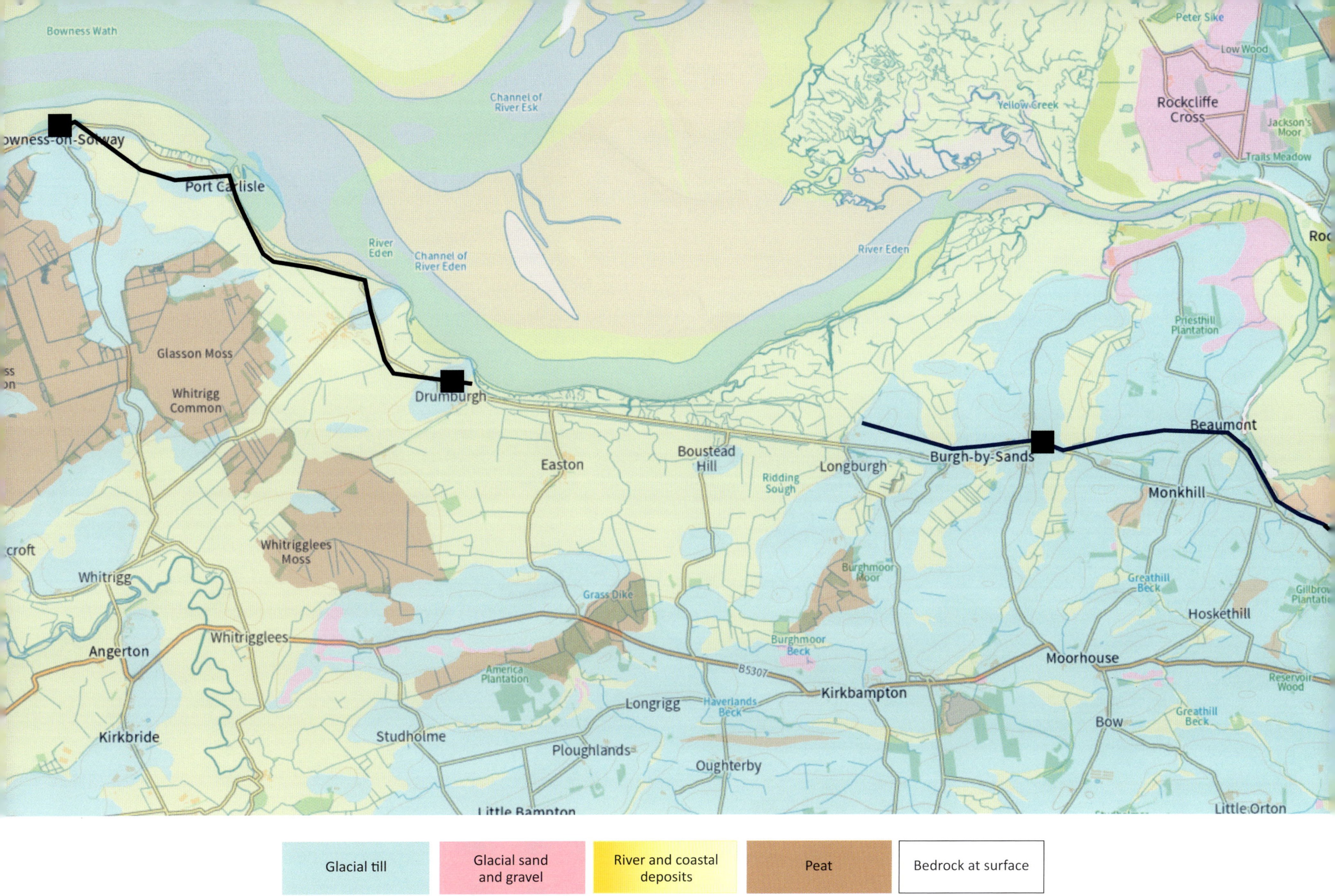

Geological map of the western section of Hadrian's Wall and forts.

Contains British Geological Survey (BGS) material © UKRI 2023.

Source: BGS Geology Viewer - https://geologyviewer.bgs.ac.uk/.

Wallsend

In terms of natural foundations, the Romans largely made rational choices about the line of their Wall all the way from the Solway to the Tyne. But their decision to extend to Wallsend was to prove problematic.

Hadrian's Wall meets the River Tyne at Newcastle (Pons Aelius). It seems like a logical eastern terminus but the Wall then continues for five kilometres to Wallsend (Segedunum). The usual interpretation of this extension is that it provided added protection for their bridge over the Tyne and against any unwanted visitors at low water. But in choosing their line east they were crossing what was to prove difficult geology.

From Byker to Segedunum beneath the soil is a veneer of fissured clay, less than two metres thick, thought to be a result of the deep freezing and thawing of the ground surface that took place in the tundra after the last ice sheet disappeared. Beneath this is laminated (finely layered) clay and silt, sediments that can be more than 40 metres thick and were deposited in the quiet conditions of a large and deep glacial lake that covered much of the Tyne and Wear area. The laminated clays are very weak and when wet become plastic and flow. Their composition and structure results in low shear strength and so unless support measures are taken in excavations, or piles are driven before building, then they will slip and fail. Deposits like this cause considerable problems for modern construction and where they occur foundations must be specially engineered.

It is no surprise that sections of the Wall just west of Segedunum collapsed and were rebuilt by the Romans not once but perhaps four times. Water is assumed to be the cause of the collapses in archaeological accounts, but water is likely to be just the trigger. The geology – the weak clays – and the load of the Wall's weight, exacerbated by sloping ground and a small valley, are the more likely root cause. The Wall foundations here were double thickness; were the Roman builders trying to mitigate a known geological hazard? Or is there no correlation and their concern was getting their barrier in place quickly not its resilience. Whichever is true one could argue that any structure that survived around 80 years on such shaky ground was a success rather than a failure.

Laminated glacial lake clays, from a borehole in nearby Walker.
Image by kind permission of C. Johnson/Roberts Environmental Ltd.

A section of the Wall is exposed west of Segedunum Fort [NZ299660]

Collapsed section of Wall at Buddle Street, Wallsend in 2015

14

Milking Gap

To the south of the Wall is the Vallum; a 6-metre-wide, 3-metre-deep, flat-bottomed ditch with earth mounds on either side. For much of its course the Vallum stays close to the Wall, but sometimes it changes direction significantly and deviates southwards.

Minor deviations at forts may simply relate to the sequencing of their construction. Elsewhere, because it is a trench, the surveyors had to be even more conscious of the geology underlying the Vallum than the Wall. Choosing the wrong course across the ground would have meant significant delays and even failure of the earthwork. Excavating the incredibly hard Whin Sill would have increased the time to complete the 130+ kilometre excavation enormously. The Vallum may have been cut through this rock at one point south of Limestone Corner but (learning lessons?) they made no attempt elsewhere and their unsuccessful efforts to complete the ditch through the dolerite north of the Wall here speaks volumes. Trying to excavate such a large trench through peat bogs and soft lake sediments would have been Sisyphean – the sides would have collapsed and the high water-table meant it would have filled with water almost immediately. Dolerite and peat appear to have been avoided if at all possible.

On detailed geological maps the Vallum's navigation of difficult ground from Sewingshields to Carvoran is obvious. It steers clear of the outcrop of the Whin Sill, finding a route in glacial clays just south of the exposed bedrock. The same geological maps show you how its excavators negotiated their way around peat bogs too. South and west of Milking Gap is a classic example: the Vallum takes a double kink to avoid thick, wet peat – an area likely to have been much wetter and perhaps a lake 1900 years ago. It takes evasive action, heads towards the Military Road (B6318) and then runs parallel to it.

The Vallum, its ditch and ramparts, would represent a huge task even for modern excavating machines. Its completion by troops equipped with only spades, hammers, chisels, barrows and carts is a staggering achievement. The purpose of all this effort? The Vallum may mark the southern limit of the military zone. There were only a few places – at forts – where it could be crossed and these causeways were closed by a gated arch. While its function is still debated, what we may deduce is that its changing course nicely demonstrates the pragmatism of its planners and diggers and their geological nous.

Take the track south from Hadrian's Wall Trail through Milking Gap [NY77677]

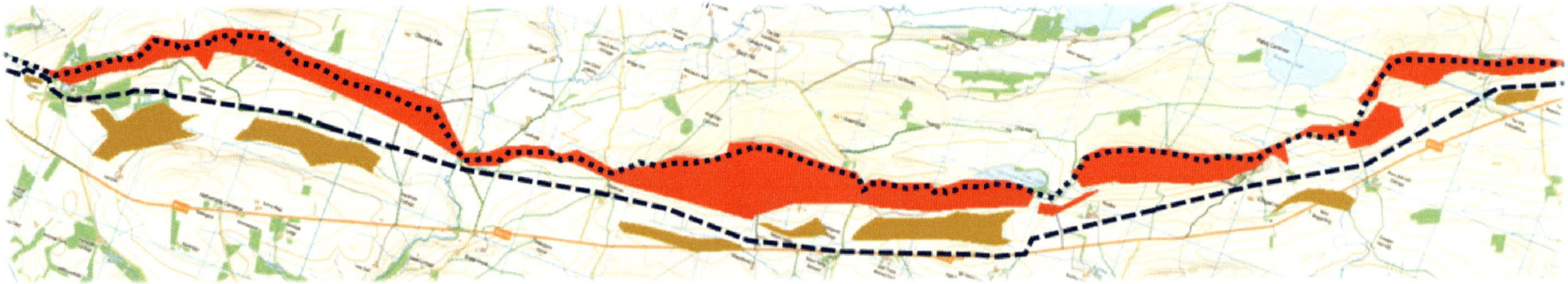

How the excavators of the Vallum (dashed line) threaded their way between Whin Sill dolerite (red) and peat and former lakes (brown)

The Vallum changes direction to avoid the peat (outlined in gold) south and west of Milking Gap.
Image kindly taken by David Barnes.

Cawfields Milecastle

Geologically, the route the Romans chose for their Wall across the Solway-Tyne isthmus makes perfect sense. The placement of some of the milecastles and turrets less so.

The terrain is the dominant influence on the line of the Wall and it is geology that shapes the terrain. In the central sector Hadrian's Wall takes advantage of the escarpment of the Whin Sill, while the sectors extending to the west and east coasts occupy the highest ground with good views south. But although the Wall heeds the shape of the ground, other elements of the frontier infrastructure appear less sensitive to it and the underlying rocks. The current thinking is that the Wall line was laid out on the ground by legionary surveyors and the positions of milecastles and turrets were staked out along this line at spacings predetermined by a plan devised remotely at a much higher level. There are 80 milecastles (double-gated fortlets with fortified gateways), one approximately every Roman mile, with two turrets between each pair of milecastles.

Milecastle 42, just east of Hole Gap and Cawfields Quarry, may be one of the most photogenic milecastles on the Wall but its position seems less than practical. It is built on a steep slope and its northern gateway is a few metres short of the north scarp of the Whin Sill. The sloping site would not have made the builders' lives easy. It does not seem like the best trade-off but one theory is that the milecastle was moved from its original staked position in Hole Gap so that its occupants could maintain visual contact with the Haltwhistle Burn Stanegate fort.

But a more fundamental factor comes into play, one that affects multiple aspects of the interpretation of the frontier's installations – the reliability of Roman Wall measurements. The standard mile is 1000 paces, fixed as 5000 Roman feet (1481 metres) in 29 BC. Along the Wall distances between milecastles vary, most widely (understandably) in the seriously undulating central zone. The Romans probably used stakes and a rope to survey and measure. As an experiment, mapping software was used to emulate the original rope measurement on just one of the 'miles'. This calculated the distance between Milecastles 41 and 42 as 1644 metres, an increase of more than 10%. Cawfields is certainly not alone in its apparent insensitivity to both the terrain and the (perceived) rigidity of the plan. Whether the builders of Hadrian's Wall were influenced by the original master plan, by an emperor's revisions, by adoption of rigid stone walls and not flexible turf and timber and not least by the terrain and its rocks, is a puzzle that is not going to be solved quickly.

Cawfields Quarry. The Wall was destroyed by quarrying the Whin Sill for roadstone in the first half of the 20th century.

Park at Cawfields Quarry, the milecastle is 350 metres east [NY716667]

Cawfields Milecastle, number 42

Willowford Bridge

The remains of the bridge at Willowford and the mile of intact Wall, milecastle and turrets to its east, are one of the highlights of the frontier.

There used to be Roman bridges across all the major rivers in the north, the Eden, North Tyne, Tyne, and Tees. Several were washed away and had to be rebuilt, multiple times. This is what is thought to have happened at the bridge over the River Irthing at Willowford and explains why the remains of the abutments and pier you can see today can be complicated to interpret. But the first and most obvious question is why is there a bridge just here when the river is a hundred metres to the west? The answer is that since the bridge was built 1900 years ago the river has inexorably changed its course, creating a typical meander loop by eroding the thick but soft, unconsolidated glacial sediments of Harrow's Scar. Rivers are dynamic and channel movement and flooding of their valleys is the rule not the exception, something our changing climate is reminding us of today.

Around AD 122 the Romans initially built a narrow bridge of stone arches, carrying a walkway (or road). There was a tower at the east end, either to control the crossing, or to provide access to the bridge. Several decades later a flood destroyed all or part of the bridge and the stone arches were replaced in timber. Around the turn of the 2nd and 3rd centuries the bridge was rebuilt again, this time with a roadway – the Military Way – and ramps to provide access. There is evidence the bridge was robbed for stone in the 4th century to repair the fort wall at Birdoswald and again in the 19th century to build Willowford Farm.

Willowford bridge abutment is worth more than a passing glance as you stride along the Wall Trail. Seek out stone channels, interpreted as a sluice for a watermill. There are large carved sockets which carried the timber arches of the second bridge. Butterfly-shaped niches are for iron and lead cramps to join the large building blocks and the narrow slots in some stones are Lewis holes – an ingenious method of lifting heavy stones by slotting in flat iron pins and using cranes. Last but not least, admire the chisel skills of the masons who shaped the huge blocks of Carboniferous sandstone. There is one stone with a large colonial coral fossil in it; 330 million years ago this animal was living in a warm coral sea on the Equator, not a cool 55 degrees north! See if you can find it.

The Wall running up to the eastern bridge abutment, now 100 metres from the River Irthing

Park at Gilsland and take the Hadrian's Wall trail west for 1 kilometre to the bridge [NY622665]

The eastern stone abutment with butterfly-shaped niches for iron cramps, bridge support sockets, stonemason's chisel marks (and a coral)

Housesteads

If you only have time to visit one place, Housesteads will show you how well the Romans understood the rocks of their frontier.

Housesteads (Vercovicium) has been described as the 'grandest station' on the Wall. Today it is the most visited Roman site. Walk a hundred metres along the path from the National Trust Visitor Centre and you get a first glimpse of the fort, a huge expanse of grey ruins dominating the skyline high on the Whin Sill ridge. Before starting the climb to the fort look at the grassy escarpment back to your left, you will see it's been nibbled. These are just a few of the hundreds of small quarries in the 330-million-year-old Carboniferous sandstone that provided the blocks for the walls of the central and eastern sectors of the frontier. Over to the northeast, 400 metres away, in a small valley – the Knag Burn – there is limestone and the site of a Roman limekiln. The path that climbs to the fort may seem steep but it's half the full 12-degrees of the grass slope; that's the angle the rocks beneath were tilted by earth movements millions of years ago.

Go through the south gate and head up to the north wall of the fort – also Hadrian's Wall. No more climbing to be done, now you can meander back downhill through the amazing array of buildings. They are typical of the forts along the frontier and feature gatehouses, towers, granaries, barrack blocks, bath house, headquarters, hospital, commander's home and most famously, the latrines. All of them and their contents are products of the raw materials the Romans sourced locally and from across their Empire. The sandstone walls and their lime mortar were from close by. The coal found in a bunker in a guard house and the lead caulking in the sandstone water tank beside the latrine were probably mined less than 35 kilometres away. Stone tanks were needed to hold rain – the Whin Sill dolerite beneath Housesteads is too compact to supply water. In the walls of the later bath house rebuilt within the fort are pieces of tufa, lightweight blocks, perhaps salvaged from an earlier bath house beside Knag Burn. Sawn pieces of tufa were seen here by the antiquarian John Hodgson around 1820 and may be local. The small museum has objects carved from the local sandstone plus quernstones and ceramic vessels imported from the continent. There are many other artefacts that make the geological connection: military and domestic iron-wear, a lead slingshot bullet, copper alloy medical instruments; and jewellery – a gold intaglio ring, jet and lithomarge beads.

But to experience stark Roman recognition of the virtues of northern rock, head just west of the fort and take advantage of the only length of the Wall you can actually walk on. There you too will appreciate the scale and dominance of their greatest geological asset – the natural rampart of the Whin Sill.

Lead caulking in a stone water tank feeding the latrines.
Image taken by kind permission of English Heritage.

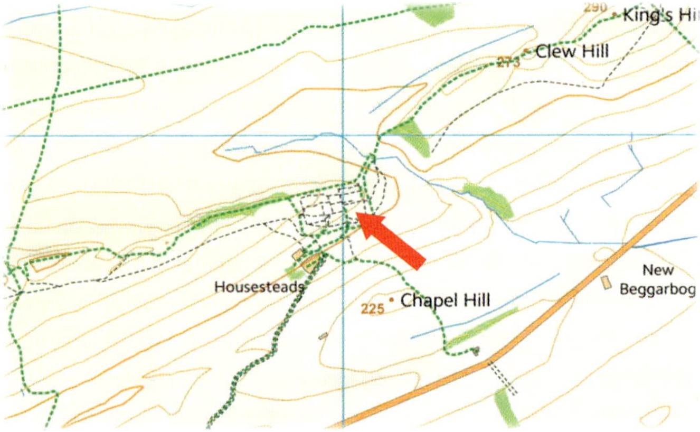

The fort, museum and visitor centre are accessible from the B6318 Military Road [NY790688]

Housesteads Fort on the Whin Sill ridge

Fallowfield Fell

To unequivocally and precisely trace where Roman Wall stone came from is an almost impossible challenge. Why?

The answer lies in how the rocks that provide the building stone originated: the conditions when they were deposited – their palaeoenvironment. On the eastern two thirds of the frontier the bedrock is Carboniferous sandstone, mostly made of quartz grains with some feldspar and mica. 330 million years ago these rocks were billions of tons of sand in the braided channels of large rivers flowing from a mountain chain far to the north. Over 20 million years the environment cycled repeatedly, river channels gave way to deltas, then lagoons, beaches and seas. These transitions left behind huge thicknesses of sand and other sediments that were constantly being redistributed. The outcome is billions of tons of rock that are rarely unique.

Head west and forward in time to the Triassic age around 250 million years ago. The red sandstones that characterise this era (and the western sector) were a product of a completely different environment: deserts with shallow transient rivers and lakes. These conditions also persisted over enormous spans of time and produced hundreds of metres thickness of fine-to-medium-grained red St Bees Sandstone. This is another rock that is virtually impossible to provenance with any precision or certainty. Geoscientists have tried; analysing both Carboniferous and Triassic rocks microscopically and geochemically and simulating past environments (palaeogeographic reconstructions). Only where there is exceptional archaeological control and the potential sediment sources are very local and distinct, has it been possible to suggest the source of Roman stones with any confidence. The frustrating outcome is that the best we can do is to conclude that most of the standard building stones in the east came from quarries as near to the Wall line as possible. In the west, where bedrock is concealed by glacial deposits, the legions had to search out more distant outcrops in the gorges of the Gelt, the Irthing and the Eden.

The quarries opened by the Romans to build their linear frontier were numerous and widespread. But the only sure way to prove a quarry to be Roman is if their carvings remain. Stone extraction took place for centuries afterwards and that, plus landscape change and erosion, usually destroys evidence of earlier working. The quarry at Fallowfield Fell, known as Written Crag, is 700 metres south of the Wall in a Carboniferous sandstone unit with a consistently thick top layer (bed). As its name suggests it had an inscribed rock – moved to Chesters Museum in 1934. It says *PETRA FLAVI CARANTINI*. Flavius Carantinus, a Roman quarryman left his name in the rock (petra is Latin for rock). All around Fallowfield are other quarries working different sandstones, their deposition separated in time from the Fallowfield stone by tens to hundreds of thousands of years. But the environments these sandstones were deposited in were largely the same and as a result the rocks are too.

A 2.5-dimensional digital model of the inscribed rock from Written Crag, now outside Chesters Museum

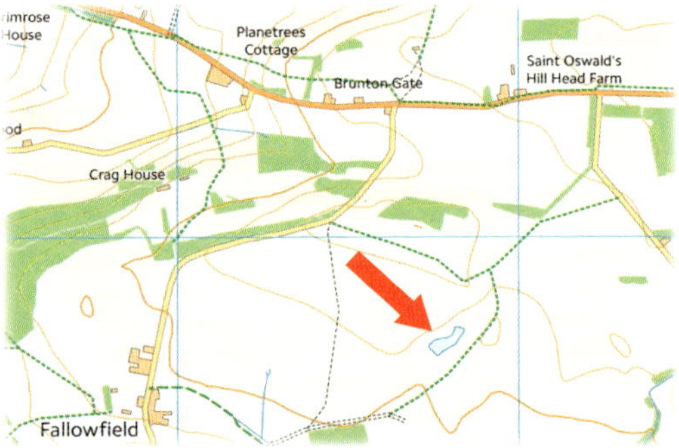

Take the minor road opposite Heavenfield and then the footpath southeast [NY937687]

The top bed of sandstone in the quarry at Written Crag

19

Combcrag Wood

The dramatic bedrock gorge at Combcrag Wood is one of several that the River Irthing has cut between its source in Kielder Forest and its confluence with the River Eden.

The sandstones, siltstones and limestones at Combcrag were once sand, silt and limey mud in Carboniferous rivers and seas deposited as sea levels rose and fell repeatedly 330 million years ago. A long period of mountain building and erosion followed, leaving the rocks bent and broken with a southerly tilt. To see that best, walk to the little bridge where Wall Burn runs into the Irthing. The bed of the stream is a creviced limestone pavement sloping south. To geologists that tilt ('dip') is apparent in the intricate multi-coloured contortions of the rocks depicted on the geological map extract shown here. It demonstrates well the fieldcraft and three-dimensional insight of the geologist who solved the geometrical puzzle of these rocks as they intersect with the convoluted topography of the gorge. Many would also see this map as a work of art.

The quarry in the thick sandstone (coloured yellow on the map) which forms the neck of the hairpin bend of the Irthing is one of very few that can be proved to have been worked by the Romans. On a hard-to-reach ledge below the footpath are seven carved inscriptions. They are important because they are not the work of skilled masons under formal instruction. These are rough carvings, cut by soldier-quarrymen as they extracted stone for the construction and repair of Hadrian's Wall and its buildings. While one, typical of all graffiti, reads: Stadus did this, another, more enigmatically, says: Apollonius; I, Daminius, did not want (to do it)!

There is an eighth inscription. It reads *FAVST ET RVF COS* and translates as Faustinus and Rufinus, Roman consuls (in AD 210). But it was proved to be a forgery, a scandalous Victorian attempt to convince people that it was not Hadrian who built the Wall, but Septimius Severus 60 years later. Such were the rivalries between antiquarians... and the kudos of ground-breaking discovery!

Roman soldier/quarrymen inscriptions

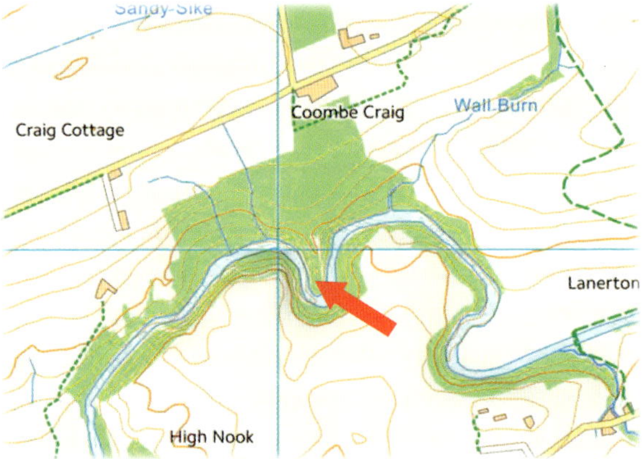

Leave Hadrian's Wall Trail at Combcrag Farm and take the footpath south through the wood for 500 metres to the River Irthing [NY591650]

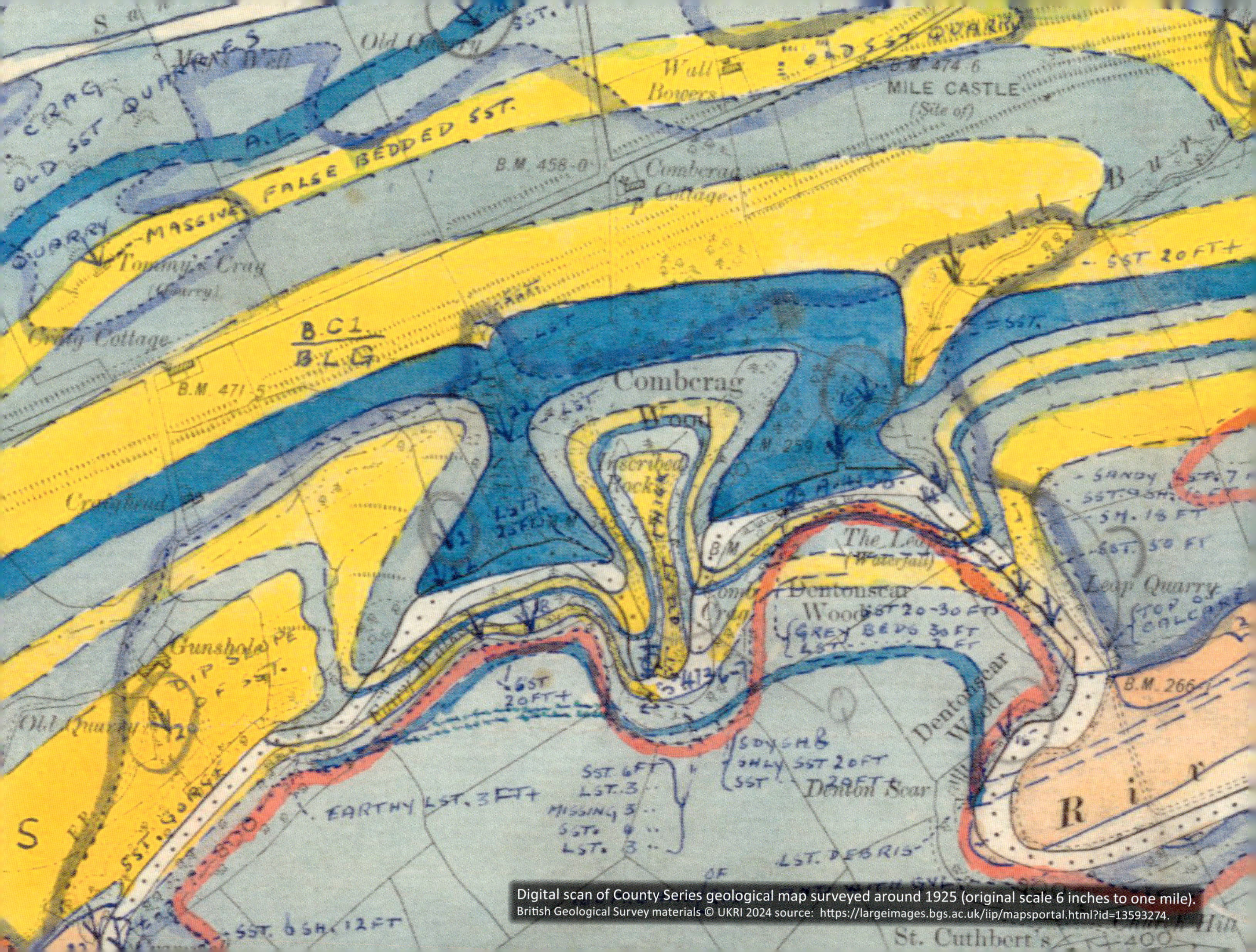

Digital scan of County Series geological map surveyed around 1925 (original scale 6 inches to one mile).

20

Wetheral

South of Wetheral, in a rocky gorge cut by the River Eden, is an old quarry – one of four in the Triassic rocks of the western sector of the Wall that have Roman carvings.

East of Lanercost the bedrock of the Wall corridor is Carboniferous in age; to the west it is largely Triassic. The dividing line between these two geological eras is not a fault, it is an unconformity and it represents more than 50 million years of erosion. The Triassic rock of the west is mostly concealed by glacial deposits, but where rivers have cut through these upper, softer sediments (like the gorge at Wetheral and the River Gelt a little further east) the bedrock is exposed. It is called the St Bees Sandstone and is around 250 million years old. It is a dull red rock with some layers and lenses of white and pale grey. It is mostly fine-grained, angular quartz but some grains are larger and rounded and so possibly wind-blown. Fossil ripple marks, rain spot pitting and the presence of thin dark brown shale layers, often 'ripped up' as pellets in the sandstone, are evidence that the rocks were deposited in shallow but transient water.

The St Bees Sandstone is an extensive and uniform body of rock, at least 150 metres thick at Wetheral and exceeding 600 metres elsewhere in Cumberland. The sandstone layers are usually horizontal with individual 'beds' up to three metres thick. Such thick beds together with the widely spaced vertical joints mean that large blocks can be quarried. This is a quality architects and city planners across the UK and the USA valued – it was exported widely in the 19th century.

It was a quality not lost on Roman masons either. Here, in the Gelt gorge and southwest of Carlisle at Shawk Beck, their carvings are direct evidence that they opened large quarries to meet their demands for stone in the west. These quarries were, of necessity, much further from the Wall than those in the east, that and their location in river gorges, would have made transport a challenge. Three inscriptions testify to the Roman origin of the Wetheral quarries. One *LEG XX VV CONDRAVSISIVS* with a stag motif, records the work of Condrausisius, a soldier of the 20th Legion Valeria Victrix. Nearby, cut into the same cliffs are three rock chambers – St Constantine's Cells – they are probably early medieval cave dwellings used for refuge from border raiders in the 14th century; hence their other name 'Wetheral Safeguard'.

St Constantine's Cells cut into the St Bees Sandstone bedrock

Take the footpath south of Wetheral on the west bank of the River Eden [NY467534]

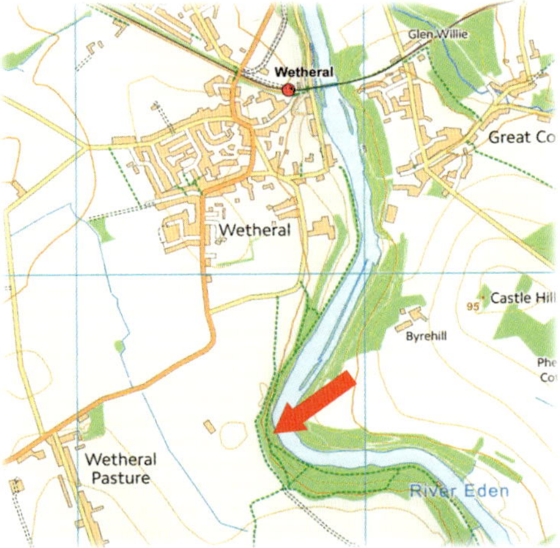

Roman inscription in St Bees Sandstone beside River Eden at Wetheral. The upper right part of the carving has split along a natural bedding plane and will soon detach.

Hardknott Fort

From its position on a rocky spur high in the Lakeland fells it controls the eponymous pass above and commands a stunning view down Eskdale. It must be the most spectacularly sited Roman fort in Britain.

Hardknott Fort defends the road that connects the forts at Ravenglass and Ambleside. Did the road carry valuable metals, ore, or even slate, to the port of Ravenglass? The geological resources of the north, like coal, lead and possibly several other minerals, surely played an important role in Roman trading. The idea that the Romans had mines in the Lake District seems irresistible but evidence remains circumstantial. So, were the road and fort simply there to exert control over potentially rebellious local tribes?

The fort is square with sides about 115 metres long, there is a gate in each side and towers in each of the four corners. The internal structures visible are typical: commander's house, the headquarters (principia) and two granaries. But the position and design of these buildings did not follow the standard plan – adjustments had to be made for the unyielding bedrock outcrops. Aqueducts fed water from nearby streams to the fort and its external four-roomed bath house. To the northeast the Romans cleared stone and constructed a huge terrace to create one of the best-preserved Roman parade grounds in the Empire. It is a levelled area of 1.2 hectares, complete with a stone platform (called a tribunal) and ramp for the commander to inspect his troops!

Hardknott Fort is built from the Ordovician volcanic rocks that underlie and surround it. These are andesites and dacites, a product of violent eruptions 455 million years ago. This Roman fortification was, like most buildings throughout history, constructed from local materials. However tough they may be to work, proximity rules. Travel around the north and you will see that localness and the diversity of our geology reflected in our architecture: the grey-brown sandstones of the Northumberland Carboniferous; the red sandstones of the Cumbrian Triassic; the grey Carboniferous limestones of the Pennines and dales; and even, where bedrock is not available, the multi-coloured cobble houses of the Solway lowlands. Hobson's Choice here in upper Eskdale were hard igneous rocks which had to be shaped, so they did. Hardknott Fort was built by the Fourth Cohort of Dalmatians (a region in Croatia) during Hadrian's reign (AD 117-138). A much smaller garrison occupied it under the next Emperor, Antoninus Pius, and it may have been abandoned around AD 150 with sporadic occupation afterwards. You won't have far to walk once you have reached the parking place below the fort but you might find the drive up or over Hardknott Pass a little more challenging!

The southwest wall of Hardknott Fort made of shaped blocks of hard ancient volcanic rock

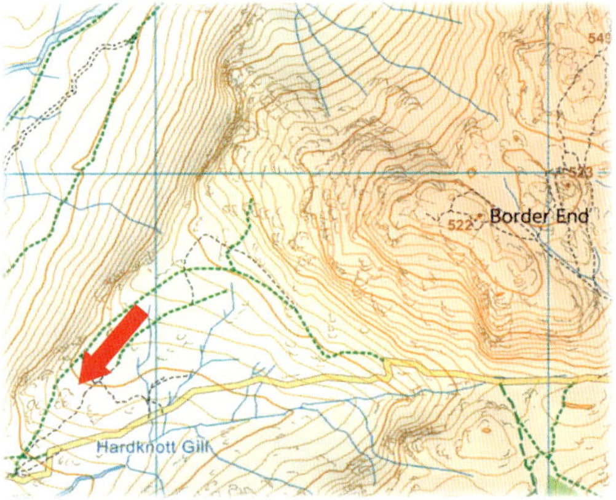

There is a small car park at Jubilee Bridge and it's a 700 metre walk northeast up to the fort [NY218014]

Hardknott Fort and Eskdale from Border End fell

Hare Hill

Whether you are an expert, or curious visitor, Hadrian's Wall can be both an enigma and a paradox.

Its remains have been studied for centuries and reading even a fraction of what has been written is a challenge. Yet the more evidence that is gathered, the more questions are raised. Just how wide and high was this wall and why does it vary. Who built it and where did they begin? How long did it take them? When did they finish? But the focus of this page is the what. What is it built of? Perhaps the most consistent characteristic of Hadrian's Stone Wall is that it has facing blocks on the north and south sides. They are predominantly sandstone, of Triassic and Permian age in the west and Carboniferous everywhere else. The foundations vary except where they are bedrock. They may or may not be set in a trench excavated below ground level. Foundation composition changes with location: a mix of blocks, flagstones, cobbles, or clay, with edge stones that are laid wider than the wall above. Between the outer facing stones, above the foundation layer for the full height of the wall, is a core of rubble, broken sandstone and/or river cobbles and near its outcrop, Whin Sill dolerite.

Because its facing stones and core have often been restored and pointed with mortar and because the Wall was so tall, most visitors assume that mortar was used when the legions first built it. But there is no archaeological consensus. Some believe the evidence shows its outer facing blocks were laid dry and others they were mortared in places but not everywhere. It is even possible that trampled clay was used instead. Bonding with lime mortar did seem to eventually become the norm. Were the initial sections perhaps experiments? The Wall rebuilt by the emperor Severus 80 years later is assumed to have used hard white mortar. That the bare grey-brown or red sandstone facing stones were once covered with white lime render is no longer regarded as likely. But the method of bonding the rubble core, sometimes with a matrix of clay and earth, sometimes mortar and sometimes all three, is still debated.

If there is a pattern in the Wall's composition it seems to point to legionary, even centurial, autonomy and a pragmatic approach of sourcing building materials locally. 1900 years of weathering and decay, destruction and burial leave an evidence base that is open to many interpretations. At Hare Hill a short section of the Wall remains, the last decent piece of surviving masonry going west. Its facing stones were re-built in Victorian times but its rubble core, mostly of angular sandstone, is said to be original and at three metres tall is one of the tallest fragments left.

The northern (restored) face of the Wall

From Banks village walk west along Hadrian's Wall Trail for 800 metres [NY564646]

The rubble core of the
Wall exposed at Hare Hill

Knag Burn

23

As you journey along the central sector of the Wall you will see many 18th and 19th century stone-built limekilns. Each one is evidence of the presence nearby of limestone close to the surface.

Limestone layers run east-west; with the other rocks they impart the grain of the country. They were once limey mud and shells in coral seas around 330 million years ago. The pressure of deep burial under thousands of metres of other sediments turned them into hard stone – essentially solid calcium carbonate, $CaCO_3$. Millions of years of mountain building episodes and erosion eventually brought them back to the surface. These rocks have provided a source of lime for centuries: for agriculture, mortar, cement and render; but also as a flux for the iron and steel industry; as an ingredient of medicines and toothpaste; and for the tanning and paper industries. To minimise the effort needed to transport the stone from the quarry the kilns are always sited close to the limestone outcrops. The chamber within the kiln was loaded with layers of limestone blocks and coal. This was burnt at a temperature of around 1000 degrees centigrade for perhaps five days and then the quicklime (calcium oxide, CaO) was raked out of the hole (the eye) at the bottom. The kilns you see are mostly from the 18th to the 19th centuries, they are called 'pot' or 'running' kilns and were designed to work with the higher temperature of coal fires and to be kept burning continuously.

The Romans used an enormous amount of lime mortar in the construction of the Wall and its related buildings and they also used lime render in fort buildings and walls. So we know they were quarrying and burning limestone. However, evidence for Roman kilns along the frontier is rare. One was reported from Vindolanda and another very recently at Rochester, but only a single definitive kiln has been documented. This was in the little valley of the Knag Burn, adjacent to a limestone outcrop a few hundred metres southeast of Housesteads Fort.

The kiln was exposed by the archaeologist F. G. Simpson in 1909. It had been created by cutting into shale bedrock and had a flue 4.5 metres long connected to an oval chamber 3 metres across. A ledge within the chamber suggests it was wood-fired. Roman pottery of the late 3rd-early 4th centuries was found overlying and within it. Although a recent archaeological study has suggested otherwise, from a geological and quarrying perspective it seems logical that the Romans would, just as they did with sandstone, extract and process limestone as close to their construction sites as possible. This suggests many kilns along the Wall. Perhaps these were ephemeral 'clamps' – hollows in the ground filled with layers of limestone and firewood covered with turf – which would leave little trace. Exactly where and how the Romans produced lime, a critical raw material for their frontier, will, without more research and clearer evidence, remain a mystery.

Site of the limekiln on the west bank of Knag Burn

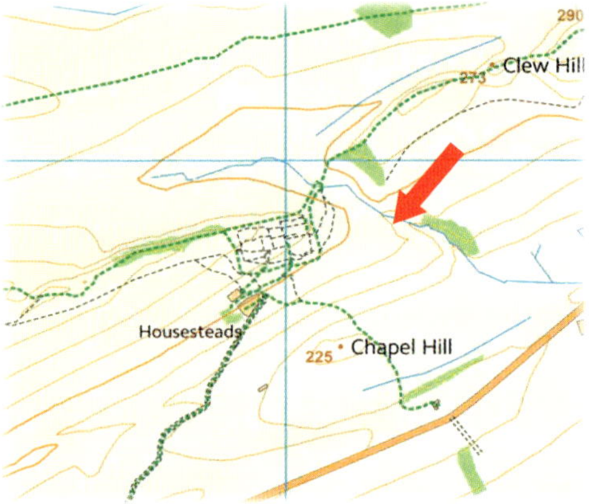

Walk east 250 metres from Housesteads Fort [NY793689]

Rare limekiln

Plan of Knag Burn limekiln excavated by F. G. Simpson in 1909.
Published in 'Watermills and Military Works on Hadrian's Wall' edited by Grace Simpson: 1976.

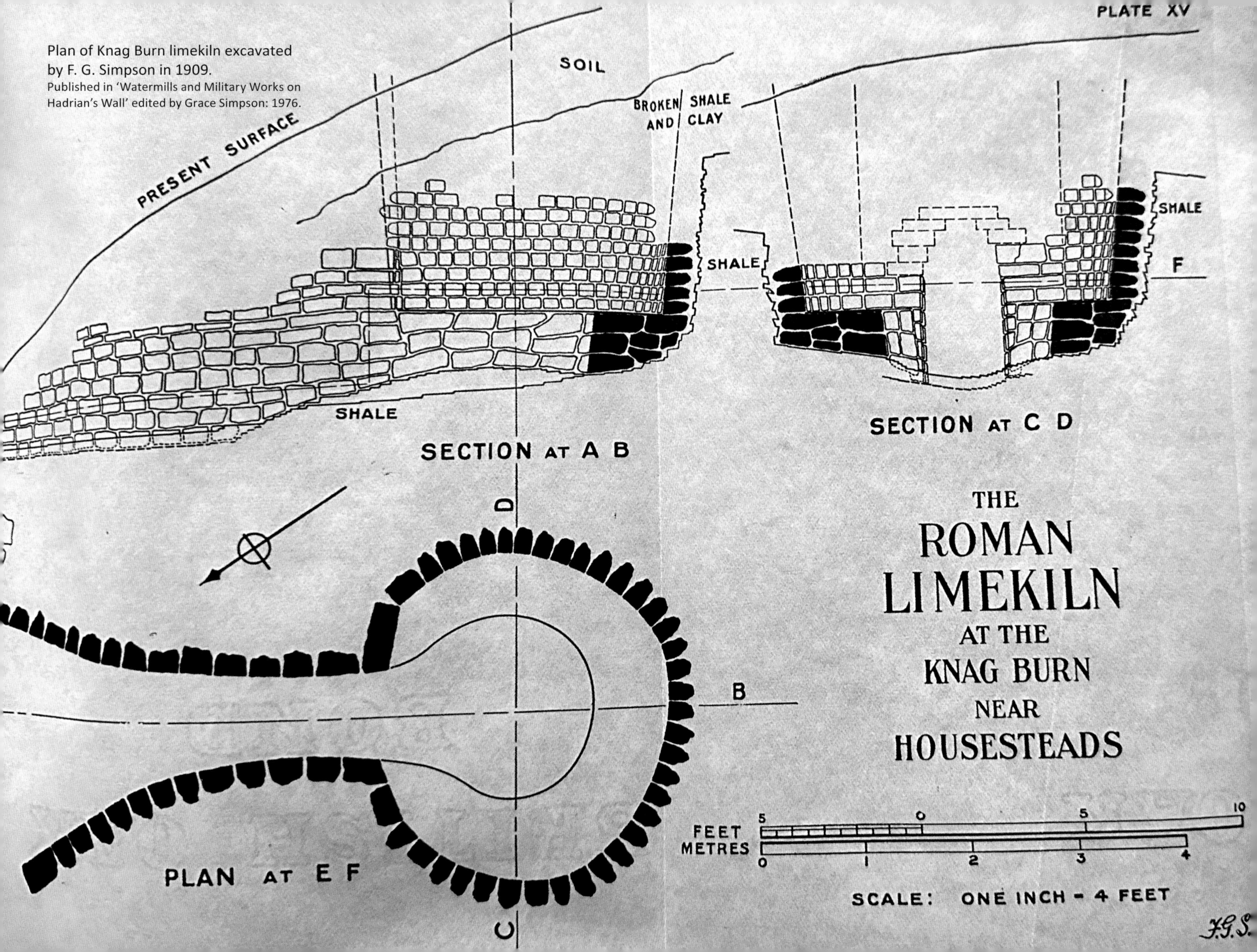

24

Newbrough

St Peter's Church at Newbrough is thought to be the site of a Roman Stanegate fort. In a field next to the church is a spring – St Mary's Well. Close by is a large block of a porous looking rock called tufa.

The word tufa is easily confused with tuff – a rock made of volcanic ash. Both can be light in weight and were used by Roman engineers to construct parts of buildings. But tufa is a freshwater limestone formed by physical and biochemical processes in streams and springs which are rich in carbon dioxide. Most tufa in Britain is very young – less than 10,000 years old. It is precipitated on bacteria, algae, mosses, plants and even manmade things. It has an open texture and can look like a rock sponge. Tufa deposits are sometimes recorded on the most detailed geological maps, but because they are usually less than 10 metres across they rarely appear on smaller scale, published maps. The detailed maps show tufa at Bewcastle, at Kirkwhelpington and a few other locations in Roman Wall country.

The Romans were masters at grand design long before modern architects and engineers, and military infrastructure on the northern frontier was usually based on tried and tested specifications. There was even a specification for bath house ceilings which required nature's equivalent of lightweight thermal blocks – tufa! Next to the bath house at Chesters Fort are several tufa blocks. Some have notches cut top and bottom. They are called 'armchair voussoirs.' The blocks of tufa were stacked together to form a vaulted roof. Flat clay tiles were then fitted into the notches to create a cavity so warm air could circulate in the roof of the bath house.

Tufa blocks and fragments have been found in other Roman excavations too: in Corbridge, South Shields (Arbeia), and Housesteads, and at Catterick where pieces were re-used as a grave cover. There will undoubtedly be other tufa pieces lying in excavated stone rubble but the rock type and its significance is not always appreciated. Where the Romans sourced the tufa found along their northern frontier is a difficult question to answer. This very young rock occurs in much larger volumes on the continent, where there are many examples of Roman use. Blocks occur in the Roman Baths at Bath and thick deposits are known in Derbyshire, Worcestershire and Yorkshire. It is possible that the Romans exploited our smaller local deposits like Newbrough, but to prove that we need to gather evidence, from possible sources and the excavations.

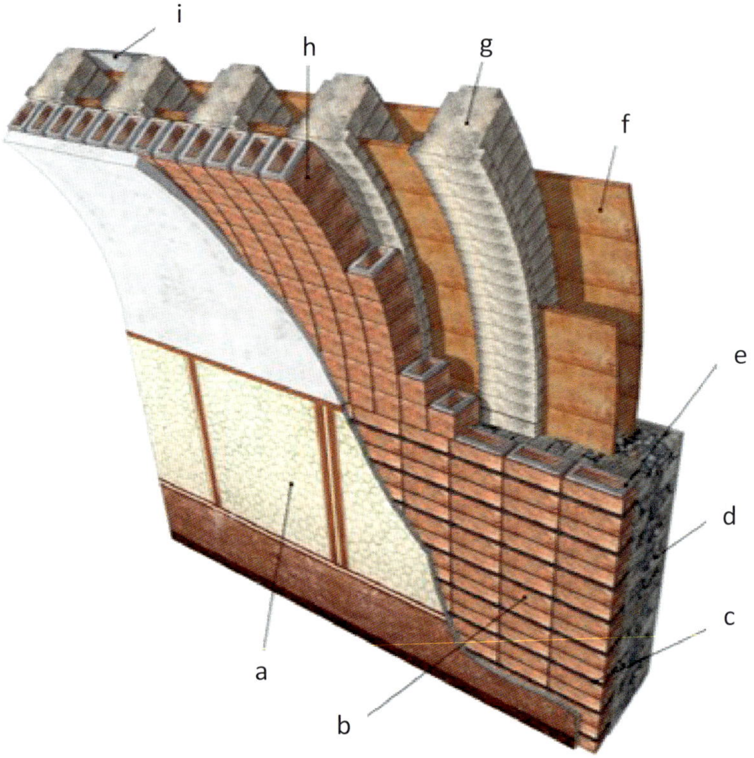

a.	wall plaster	d.	exterior wall	g.	armchair voussoirs
b.	box tiles	e.	mortar	h.	hollow voussoirs
c.	iron hold fasts	f.	tiles/slabs	i.	concrete

Section of a bath house wall and ceiling.
Image from Roman County Durham (2021)
by kind permission David Mason © Durham County Council.

From the church take the footpath to the northeast for 250 metres [NY870681]

Block of tufa beside Newbrough Burn.
Inset image: Armchair voussoirs beside the bath house at Chesters Roman Fort.
Image taken by kind permission of English Heritage.

Cotehill

25

They used it to encase bodies for burial in York and mined crystals in Spain big enough to glaze windows in Pompei. It is naturally abundant in Cumbria. But did the Romans use it along the frontier?

Gypsum ($CaSO_4.2H_2O$) is hydrated calcium sulphate. As well as producing the pink wall-plaster we all know, it has other, rarer, forms: alabaster, used in sculptures and carvings; and selenite, the translucent crystals used for the windows in Pompei. In the north of England gypsum formed around 270 million years ago, when we were just 15 degrees north the Equator. Our environment at that time was transient lakes and warm seas that, as they evaporated, precipitated minerals like gypsum and salt. If you crush and heat gypsum and drive off two thirds of its water you create plaster of Paris; re-wet it and it will set rock hard. Mixed with sand and other ground rocks gypsum makes wall-plaster. Its uses have been many: from creating fine monuments; to whiteners and fillers in paper and paint manufacture (it was known as Terra Alba); to fertiliser and suppressing dust in coal mines; and even sometimes to adulterate flour!

Near Cotehill, eight kilometres southeast of Carlisle, gypsum was quarried and mined from the 17th century until 1966. Ponds and old rail sidings are evidence of what was once a thriving Cumbrian industry. While gypsum dissolves very quickly in water and outcrops weather away, the thick layers here, beside the River Eden near Long Meg and at Barrowmouth south of Whitehaven, were once visible. That means they may have been noticed by the Romans and potentially exploited. We know that they understood the properties of gypsum from its use elsewhere in the Empire and because it was used by Egyptian and Greek civilisations before them. However, there is an abundance of limestone in our part of the world and it seems likely that this, coupled with the fact that lime was the recommended ingredient for plaster and render by Roman builders, means gypsum may have been a local natural resource that went untapped.

This was not true a little further south. In and around York at least 45 Roman burials have been found which underwent an intriguing ritual: coating the shrouded bodies in liquid gypsum, or its derivative, plaster of Paris. The source of the gypsum is not yet known and the practice is unusual but does occur elsewhere in Europe and Africa and may be restricted to high status individuals. A recent programme of three-dimensional scanning of the casts has revealed that the gypsum is preserving fascinating details that were previously unknown.

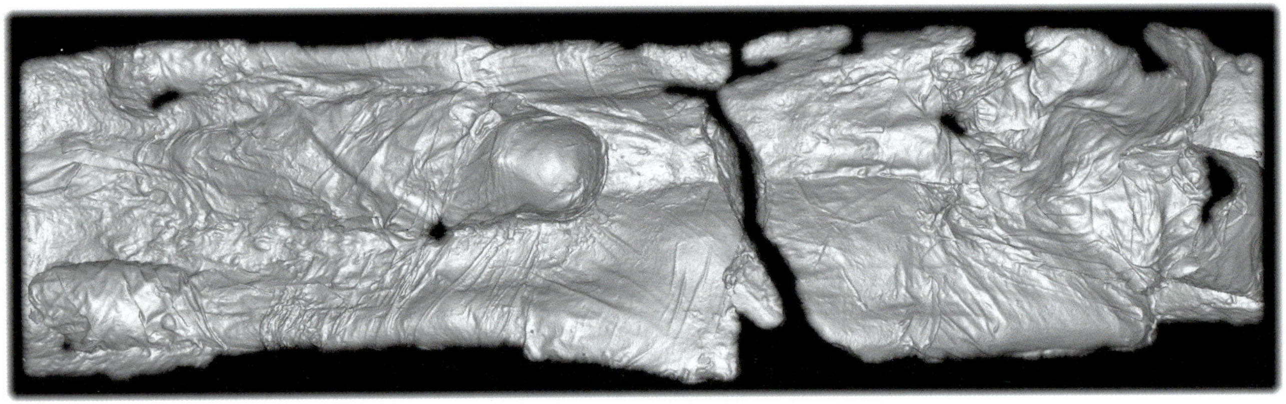

Three-dimensional scan of 3 Roman bodies found in York encased in gypsum/plaster of Paris.
Photograph courtesy of Maureen Carroll.
Image by Heritage360, University of York.

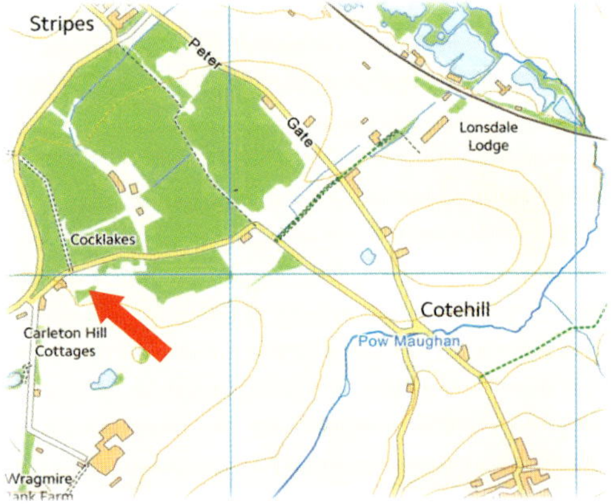

Cocklakes is to the northwest of Cotehill. The gypsum mines in this area were reinstated but the rail lines that serviced them are still visible [NY457511]

Versatile but ignored?

Cocklakes gypsum mine in 1927. Gypsum and anhydrite form the white layer, which is 6 metres thick.
Inset image: the restored mine site today today.

26

Scalesceugh

1900 years ago, at Scalesceugh beside the ancient road from York, clay pits and kilns were producing tiles to roof buildings in Roman Carlisle.

The large red-brown clay roofing tiles found here and in Carlisle are stamped *LEG XX VV* and *LEG.VIIII H*. The work was being done by two legions: the 20th Valeria Victrix and the famous 'lost' Ninth. You can see examples in Tullie House Museum, Carlisle. Excavations and geophysical surveys at Scalesceugh have revealed the remains of at least 25 kilns. Why were the kilns here? The Romans were exploiting an ideal raw material – a small deposit of stone-free laminated clay less than half a square kilometre in area. Perhaps soldiers from the nearby fort at Park House, Wreay accidentally stumbled on the clays exposed in the banks of the River Petteril, or maybe the ceramic potential of this soft, stone-free clay became apparent when they built their road across it. Or perhaps local tribes were already aware of its qualities?

The clays have a geological story to tell too. They were once very fine silt and mud laid down as delicate layers in an ice-walled glacial lake as the climate warmed and ice melted around 15,000 years ago. Such lakes were common in the Carlisle Plain at this time and there are numerous deposits of laminated clay. Similar clays were worked at Botcherby and Upperby in the 19th century and possibly well before and there was a Roman tilery and pottery at Brampton. Up and downstream of the laminated clay deposit at Scalesceugh, and across a much wider area around Carlisle and Brampton, are extensive spreads of sand and gravel. They were once the bed load of turbulent glacial rivers which flowed under and within the same wasting icefield that covered thousands of square kilometres. Even more widespread across the whole region are tough stony clays (boulder clay, till or diamicton), deposited directly by the ice as it streamed across the landscape and then melted.

Deciphering exactly what happened as multiple ice sheets advanced, stagnated and decayed across the Carlisle Plain in the last 40,000 years is a complicated challenge. What was the sequence of events, where was the ice from and which way was it moving and when? The different strands of evidence – field mapping, LiDAR, sedimentology, borehole logs and Carbon-14 dating – are patchy and often contradictory. Sounds just like archaeology, doesn't it?

Clay pit at Botcherby working laminated glacial clay in 1922.
© UKRI 2024. 1922 Image Ref: P202236.
Available: http://geoscenic.bgs.ac.uk/asset-bank/action/viewAsset?id=53009&index=0&total=2&view=viewSearchItem.

The clay deposit and kilns are in the field south of the footpath that runs past Scalesceugh Hall [NY448495]

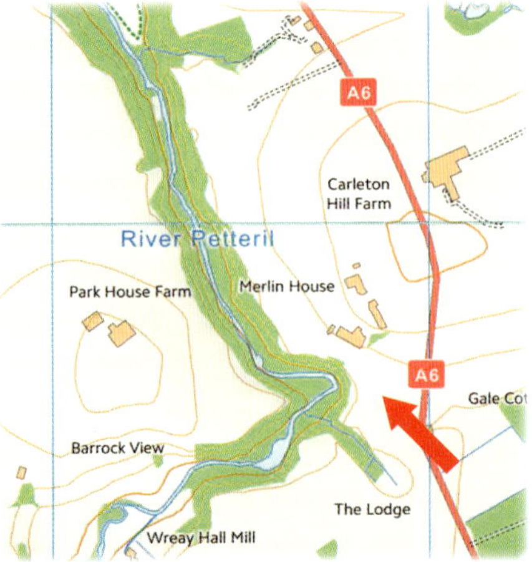

Broken roof tile with the stamp of the Ninth Legion. Found in excavations at Scalesceugh Hall, now in Tullie House Museum.
Image taken by kind permission Tullie House Museum and Art Gallery Trust.

Corbridge

Sometimes its line is obvious on the ground, sometimes modern roads follow its course, but mostly the Stanegate is no more than a dotted line on an Ordnance Survey map.

Literally the 'stone road', the Stanegate runs for 41 Roman miles (61 kilometres) between the towns of Corbridge and Carlisle. Stanegate is not its Roman name, it is medieval, but it describes well what it's made of. Most excavations of the road were done more than 70 years ago: at Corbridge, at Walwick just west of the North Tyne and near Vindolanda and Haltwhistle. In 1956 and 2022 water main trenches intercepted the road west of Newbrough and near Settlingstones. Apart from a lack of marginal ditches, the Stanegate is regarded as a classic Roman road. It often has sandstone kerbstones, and between them layers of rounded cobbles, or broken, angular, sandstone and occasionally limestone and sometimes clay. Sandy gravel makes up the top (wearing) course and can be interbedded with the other layers. It has been measured at almost 7.5 metres wide and half a metre high in places. The materials used in its construction, especially the limestone, point to local geological sources.

Roman Corbridge (Coria) is the eastern terminus of the road and its junction with Dere Street. Coria (or Corstopitum) is a very well-preserved Roman town with elegant buildings and monumental granaries. A good section of the road is visible as the main east-west street which runs past them. The exposed surface is gravel from the River Tyne: cobbles of Carboniferous sandstone with a few glacial erratics of Silurian and Ordovician metamorphic and volcanic rocks. You may not be able to see the material the road is made of elsewhere but it is sometimes visible in the landscape as a low linear mound with a noticeable camber. It used to have milestones but today only one and a stump survive – in a field just northeast of Vindolanda and the fragment a Roman mile up the modern road to the west.

The date the Stanegate was built is disputed. Some think in Agricola's time, soon after the Romans moved into northern England around 70AD. Others argue it was constructed under Trajan's rule 20 years later because its course changes to connect pre-existing forts and fortlets along its length. It links places like Nether Denton, Throp and Old Church Brampton, rather than taking a direct route over higher, easier ground. Whatever its date, the road is still viewed as effectively marking the frontier of the Empire until Hadrian built his Wall in AD 122. The Stanegate stayed in use throughout Hadrian's reign and when the next Emperor, Antoninus Pius, moved the frontier north to Scotland it remained a routeway. It was a highway until the end of the Empire and in places still is today.

The Stanegate heading east to Vindolanda and beyond, with the Iron Age hillfort and 1st century Roman signal station on top of distant Barcombe Hill

The Roman Town is an English Heritage site to the west of Corbridge [NY982649]

The main street and Stanegate in Corbridge Roman town.
Image taken by kind permission of English Heritage.

Epiacum

Seen from above, the multiple rhomboid-shaped defences of this Pennine fort three kilometres northwest of Alston are striking.

Whitley Castle sits in dramatic isolation guarding an ancient road, the Maiden Way. Now known by its probable Roman name, Epiacum, it is the highest stone-built and possibly the best-preserved fort in Britain. However, this fort may have another reason for its location: Roman state protection of the mining and processing of lead (and silver?) ore in the North Pennines, a region that 1600 years later became the largest source of lead in Britain. If the Maiden Way was merely a road joining places to the north and south, more effective courses could have been chosen: a lowland route via Carlisle, or due north from the Durham dales. The inference is that the road, and thus Epiacum had another purpose – securing the extraction of minerals. Hike the Maiden Way south of the fort and you will see fragments of lead ore (galena) amongst the stones of the track; there are many galena veins a few kilometres south and east of Epiacum.

At the Roman fort at Brough on Stainmore Pass, lead seals were found that were used to identify packages. Several of the seals had the stamp of the Second Cohort of Nervians who spent time at Epiacum. A few seals were marked *metal*, Latin for mines. Could these be labels on consignments of North Pennines lead ingots (pigs) destined for York, the south and the continent? Other finds of lead-related artefacts, for example at Corbridge, may indicate a northern outlet for Pennine mining. Lead pigs with Roman inscriptions have been found near possible mines in Yorkshire and Derbyshire and it is probable that lead was mined by Iron Age peoples before the conquest. There is support in ancient literature too. Pliny the Elder describes Britain as having abundant mineral wealth and Tacitus reports Agricola saying to his men that among the prizes of the invasion there would be gold, silver and other *metals*.

But evidence for Epiacum being the Roman centre of significant lead mining and processing in the North Pennines remains circumstantial. An early 19th century excavation of the bath house floor revealed mortar containing 'lead mine spar' (crystals associated with veins of galena). Two large lumps of galena were found at the fort in the 1960s. Despite being assayed in the hope of establishing a local (and ancient) mining source, the heavy weathering of these lumps shows they have lain at the surface for many thousands of years and so their origin is equivocal. Crucially, in 2007-8 comprehensive English Heritage field and geophysical surveys on the site found no sign of lead, its mining, or its processing. Nevertheless, while no lead ingots or mines have been discovered, to many people the indirect signs of lead exploitation in the North Pennines are compelling. But related, oft-repeated assertions that it was silver in the lead ores of these hills that was a prime reason for the Roman presence are a lot less convincing. Available geological data does not support that.

One of the two lumps of galena found at Whitley Castle in the 1960's.
Image taken by kind permission Alston Moor Historical Society.

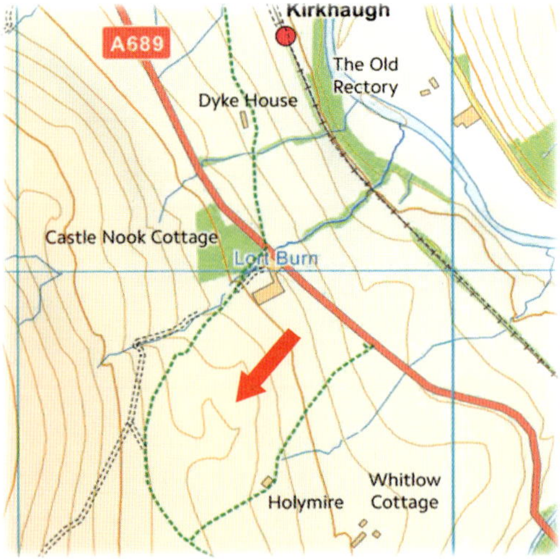

Park at The Nook (Epiacum Café). The fort is a 400 metre walk southwest [NY695487].

A mining outpost?

Epiacum from the west.
Image kindly taken by David Barnes.

Tynehead

There is something romantic about precious metals that seduces even the most erudite researcher. Are stories of large quantities of silver-rich lead ore in the North Pennines veins a myth or reality? What do the data say?

The different ways silver content in rocks was measured doesn't make an evaluation easy but the bottom line is that across the North Pennines the percentage of silver present is very low. The orefield is silver-poor not silver-rich. Between 1729 and 1870 the two major companies mining and processing lead in the area recovered 5,450,000 ounces (around 150 tons) of silver from 1,542,184 tons of lead concentrate. That is an average of only 3.5 ounces per ton of concentrate. The majority of 32 individual mines in the district recorded less than 8 ounces per ton of lead metal, but three mines, Clargillhead, Stow Crag and Sir John's show much higher levels. Sir John's Mine, (a Site of Special Scientific Interest) is an old lead mine 200 metres south of Tynehead. It was not the most productive of 19th century lead mines, but one of the veins mined recorded 40 ounces of silver per ton of lead metal. About a kilometre north is a field (curiously known as the Chesters) on alluvial gravel pock-marked with diggings. Silver levels in galena reported there were also relatively high and some have linked these workings to the Romans.

Elsewhere in the North Pennines older and unverifiable historical accounts report a mining boom and outputs of up to 4 tons of silver per year during the 12th century. If these are true, they suggest that in places richer silver-bearing lead ores may have been more accessible then than known in later periods. Perhaps they were mined from deposits nearer the surface and it was these shallower, more argentiferous lead ores that the Romans worked. Why might they be richer in silver? Like the other mineral veins across the Pennines their primary origin was deep, warm mineral-rich fluids circulating 300 million years ago. Theoretically, at least, millions of years later nearer-surface water percolating through the veins could have enhanced silver levels further.

If we accept that richer ores did occur in some places and that they were accessible to early miners, it nonetheless remains true that across the whole orefield the amounts of recoverable silver in galena are extremely modest. There are those who argue that normal mineral economics do not apply when Roman labour (soldier/slave/prison) may have been free and ores with very small concentrations of silver could have been processed. Or that the lead may have been de-silvered to improve its quality. These and other hypothetical societal factors lead some historical and archaeological accounts to imply that silver in the North Pennine hills was as important as lead to the Romans and a prime reason for their interest. For most geologists this pushes the existing evidence too far. Of course, whatever the scientific facts of the silver resource are, the Romans could have been seduced north by tantalising tales of fabulous riches in remote mountains. Many of us are.

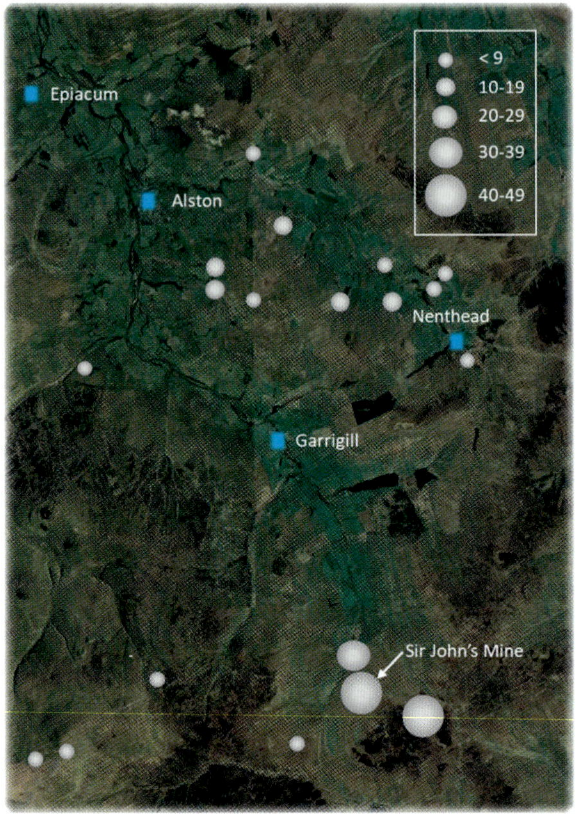

Ounces of silver per ton of lead metal in mines in the Alston mining field

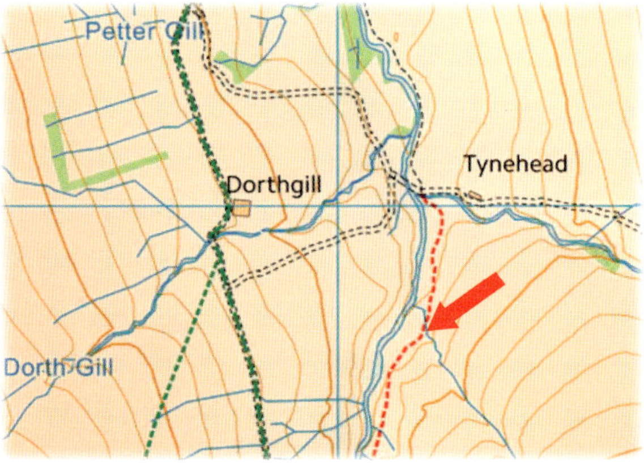

Take the no-through road south from Garrigill. Park at road end and walk to Sir John's Mine via the track to Tynehead [NY761378].

Sir John's Mine

Botchergate, Carlisle

Until the 1970s and the opening of the M6 Botchergate was a busy thoroughfare. The road goes back to Roman times, but it wasn't lined with shops then; beside it was a large cemetery and after that an industrial area believed to be smelting lead.

Carlisle originated in AD72-3 as a Roman fort (Luguvalium) located where the medieval castle is today. Archaeological investigations are revealing it as a town of considerable importance. It was a major hub for the Roman army and a prosperous administrative centre for the 'Civitas Carvetiorum' a self-governing British tribal group. In 2001, ahead of redevelopment, archaeologists excavated a site in Botchergate. In the layers above burials and cremations was substantial evidence of the smelting and working of lead. In addition to many pieces of slag there were fragments of galena ore, a lump of lead and the remains of furnaces and probable smelters. In the context of the town the location of smelting downwind on its south and east margin makes sense – turning ore into lead is a toxic business. Unlike smithing which can be done in any settlement, smelting is a significantly more complex and demanding process.

Normally metal ores like galena are smelted nearer to the mine but Carlisle is at least 20 kilometres from the closest lead veins. Did the state and military imperatives related to smelting lead outweigh the effort of transporting heavy ore? Modern analysis of galena and lead samples gave a relatively high silver content (25 ounces per ton of galena) and indicate affinities with ores from both the northern Lake District and Tynehead in the North Pennines – either, or both, perhaps the elusive Carlisle Silver Mines in medieval references. The North Pennines, while further away, is viewed as the more likely source of the galena but more analysis is needed. The higher silver values prompt several questions. Including, for example, why there is no sign of separating silver from the galena (called cupellation or desilvering), even though there was evidence of secondary working of lead here.

Lead (plumbum, Pb) was probably the most widely used metal in ancient Rome. It is soft and malleable and has a low melting point. Transported as ingots (pigs) it was turned into sheets and water pipes, jars and tableware. It was added to coins and paint, it formed roofing and coffins, shrines and ornaments. Water cisterns and pots were made and sealed with it. Romans used lead in skin-bleaching aids and in cosmetics. This meant their food, water and the things they placed on their body all contained high and unhealthy levels of lead, but perhaps its most deadly use was as slingshot bullets. Lead was recycled multiple times, that and many variables related to its geological formation, goes some way to explaining why so often lead defies attempts to trace its origin.

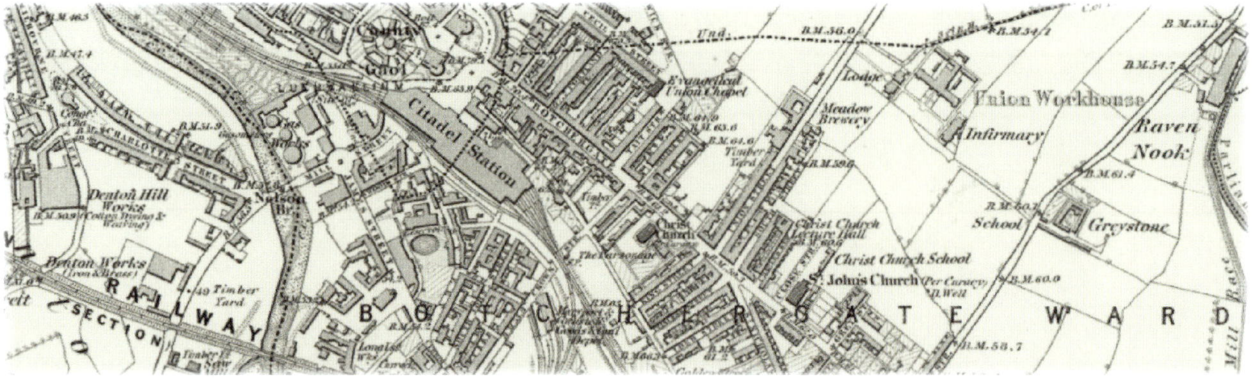

Old Ordnance Survey 6 inches to 1 mile scale map of Botchergate area, Carlisle

On the northeast side of Botchergate, Now the Ibis hotel [NY404555]

Clay base of the probable lead smelting furnace in Botchergate with the run-off channels.
Image by kind permission of Oxford Archaeology.

Burnswark

Burnswark is a spectacular table top hill about 18 kilometres north of the border. Its distinctive shape means you can recognise it from Hadrian's Wall.

It is a classic mesa, its form the result of a cap of harder 350-million-year-old basalt lavas protecting softer older sedimentary rocks beneath. But it is not only the topography and its rocks that impress, it is the dramatic narrative woven around its archaeology. Recent investigations have produced evidence claiming to show that it was the location of an assault by the Roman Army on indigenous Iron Age people. Burnswark was a large hill fort covering seven hectares. There are traces of a ruined stone wall on the edge of the escarpment and of an outer rampart on the south side. Immediately to the north and south of the hill fort are the ramparts of two unusual Roman army camps.

Several hundred lead slingshot bullets and some ballista balls have been discovered in and around the fort and the camps. The camps have been interpreted as actual siege works, but this was challenged and an alternative explanation put forward. Many slingshot bullets were on top of the Iron Age rampart which had already collapsed and others found in locations unrelated to any assault. The conclusion: Burnswark was a training range. But new investigations – metal detection, isotope analysis and experimental archaeology – have provided further evidence and reignited the controversy.

The slingshot bullets are a significant part of the debate. While superficially similar they take different forms, some are like acorns and others like back-to-back cones. Some have small holes in them and have been shown to emit a frightening whistle as they travel to their target. Where the lead came from remains uncertain. Isotopic analysis can identify its geochemical composition and whether the bullets and other lead objects have the same signature and origin but establishing where their source is much less straightforward.

The problem is that there is overlap and isotopic variation within and between these sources and also there are insufficient data; so there is less confidence in the results. This means that statements about their origin being the Pennines, or the Lake District, or more distant sources, like the Mendips, or Wales must remain tentative. The geological variables in lead genesis means the bullets could be a mix from one vein or several ore deposits, or it could be as mundane as some early scrap metal recycling – the product of re-melting lead objects like water pipes or coffin sheets. More geological input on minerals, metals and mining in the frontier zone would undoubtedly help archaeological research in developing science-based interpretations.

Lead slingshot bullet found at Housesteads and on display in the museum.
Image taken by kind permission of English Heritage.

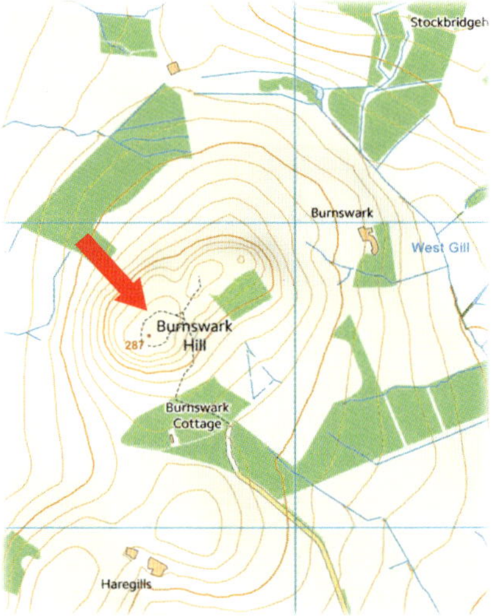

Take the Ecclefechan turn-off on the A74(M) and then go four kilometres along a minor road to Burnswark Hill. There is a small place to park and it's a 500 metre uphill walk from there. [NY186786].

Burnswark looking east, with the ramparts of the Iron Age hill fort and the two Roman camps.
Image kindly taken by David Barnes.

Brackies Burn

We don't know if, or where, the Romans mined iron here in the North but we do know they used a great deal of it.

Iron slag, plus the fragments of the fuels that may have been used to heat the ore, are often found. But whether this is from smelting (the production of metal from ore) or from smithing (the working and reworking of iron to make things) is difficult to determine – costly analysis and expertise are not always available. Establishing whether ancient mining took place is a challenge that is not unique to iron – older mineral workings almost always get obliterated by younger ones. But Roman iron mining and production in the Weald, the Forest of Dean and around Northampton is well documented. Were these the sources of the iron that was then transported long distances and turned into so many military and domestic objects all along the frontier? Or did the Romans use local 'bog ore' – a secondary and highly variable iron deposit formed since the last glaciation in ill-drained ground by the biogeochemical oxidation of iron-rich water.

Based on research some distance from the Wall it is often asserted that it was bog ore that Iron Age peoples and Romans used here. But, with rare exceptions, that remote evidence is circumstantial and where primary ore sources are available, as they are along the frontier, this presumption would be unsafe without site-specific confirmation. It is equally possible that both societies exploited local geological sources of iron in the rocks. One resource is the extensive deposits of Carboniferous iron ores called siderite (iron carbonate - $FeCO_3$) which occur as clay-ironstone bands, or nodules (called cat's heads and ballstones) in shales. We know they were mined here for many years, most extensively between the 18th and 20th centuries, for example, at Bellingham, Redesdale, Shotley Bridge and Consett.

There are many other places across the North where workable deposits of siderite ore are found and they occur near the Wall at quarries near Chesterwood and on Cawfields Common. There is a source in a shale (mudstone) above the Three Yard Limestone beside Brackies Burn less than 300 metres northwest of Vindolanda. Vindolanda is a fort with *fabricae* – workshops – and clear evidence of industrial and iron working activity. The iron ore at Brackies Burn is typical – heavy siderite nodules in the shale. Uneven ground on the north side of the burn is the product of extensive 'modern' workings, probably from the 18th and 19th centuries. There are obvious natural exposures of the shale and nodules in the meander scars of the burn and these would have been clear to the Romans. We know the Roman army worked sandstone, limestone and possibly coal beside Vindolanda. Throughout history miners and quarrymen, with the generic skills and the investment in infrastructure, invariably extract all other worthwhile resources that exist at a location. Did they exploit these local resources of ironstone too, or use bog ore, or rely on southern iron imports. Or perhaps, over 320 years of occupation, used all three?

An ironstone nodule from Brackies Burn

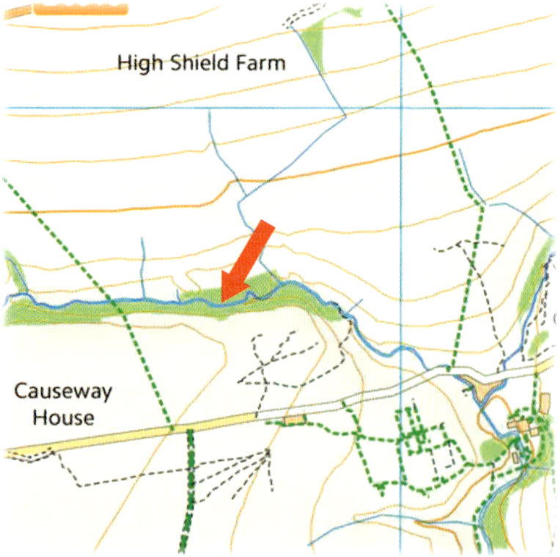

The exposures of the shale with the ironstone nodules are in the Burn to the east of the footpath [NY766666]

A layer of ironstone nodules in Carboniferous shale in a meander scar in Brackies Burn

Eskdale

33

Search the literature for places in the north of England where iron has been proven to be mined by ancient societies – Iron Age people and the Romans – and you won't find any.

If on the other hand you were to look for places where iron ore resources occur, you would end up with a sizeable list. Iron is a very common element in the Earth's crust. It is usually widely dispersed as tiny particles but when concentrated it becomes an ore. Initially minerals become concentrated because of the direct and indirect effects of tectonic forces and the intrusion of molten magma. Northern England experienced not one but two major events that formed minerals and metal ores. The first happened around 400-500 million years ago in the Lake District. The second, about 300 million years ago, the same time as the Whin Sill was being intruded, affected the rocks of the Pennines and Cumbria. Huge granite masses deep in the crust are linked to both events. It is their heat that drove the circulation of warm mineral-rich fluids through surrounding rocks. Secondary processes followed. Initially there could be erosion and sedimentary action by seas and rivers which may have redistributed and concentrated the minerals. Later, fluids moved the minerals around in the rocks and concentrated them further, making them more valuable.

Iron ore can take many forms – hematite (Fe_2O_3), siderite ($FeCO_3$), magnetite (Fe_3O_4), limonite ($FeO(OH){\cdot}nH_2O$) and goethite (FeO_2H). These ores can occur as crystals; in veins and 'flats' (layers) and as nodules. All these ores can be smelted to produce iron but the most productive iron resources in the north of England are the hematite deposits in the Carboniferous sedimentary rocks of southwest Cumbria. There is no definitive proof that the Romans mined hematite there, however, local mining experts think it is inconceivable that Romans did not notice these striking deep red and dark grey ores and exploit them. They also argue that some of the slag that has been found here is Roman, but until more research is done this must remain conjecture.

Hematite was mined on a large scale from the sedimentary rocks of southwest Cumbria 100 years ago but there are no places where these deposits can be easily seen today. However, if you are willing to take a short but rough and steep hike above Boot in Eskdale, a related hematite vein and its associated 19th century mine workings cutting through granite, occur in Nab Gill. If that sounds too much like hard work take a ride on the Ratty Railway instead. It's very relevant to the topic as it was built initially to take the iron ore from Boot to Ravenglass – site of a Roman fort – and it will give you some stunning views of the Eskdale valley.

Scree and spoil from hematite mining in Nab Gill

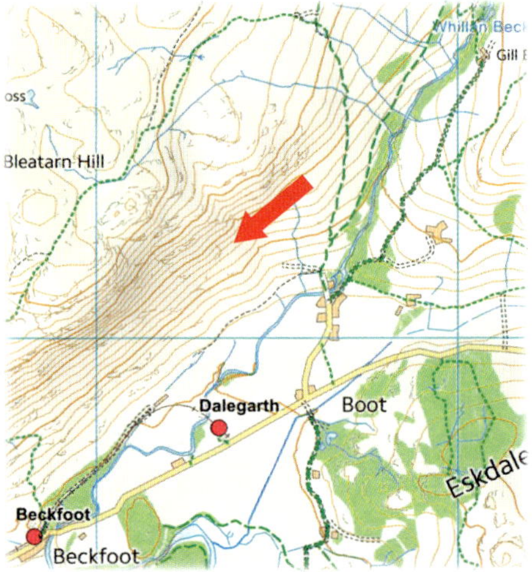

Take the track beside the Gill northwest of The Boot Inn [NY174013]

Hematite in the vein at Nab Gill, above Boot, Eskdale

Middleton Tyas

Visit Roman sites and museums and the display cases will be full of metal artefacts. Brooches, bracelets, statuettes, spoons, coins, locks, domestic vessels and even surgical instruments.

The composition of most of these will be described as copper alloy. The main metal is copper, in variable amounts, but on average around 80% of the alloy. Mix copper with tin and it produces bronze; with zinc, brass; and with both, gunmetal. Lead can sometimes be added too. Roman smiths were skilled in alloying and reusing metals. The metal mixes they made reflected not only a wish to achieve qualities like hardness and workability but the purpose and symbolism of the objects too. Copper alloy items were made in workshops along the frontier, but where did the copper they used originate? The only hard proof of Roman (and Bronze Age) copper mining is found much further south, near Llandudno and Machynlleth, Wales and at Alderley Edge, in Cheshire.

The assumption has been that the Romans transported copper from these places. But there are copper minerals in rocks of the Pennines and in the Lake District. At Coniston ore has been heavily worked since at least the 16th century, but as in the Pennines there is no sign of prior Roman working. On the Solway coast of Dumfries and Galloway there is evidence of post-medieval copper mining and circumstantial indications of prehistoric and perhaps Roman copper mining. An 18th century parish record referred to ingots of copper and a late prehistoric or Roman 'bun ingot' has also been found.

At Middleton Tyas, beside Scotch Corner, there was a very attractive local alternative. Copper was being mined extensively here in the 15th century and the spoil from many 18th century mine shafts can still be seen. The ore, which is extremely rich in copper (around 65%), originated as mineral-rich fluids which crystallised in the Carboniferous limestone and sandstone, mostly as veins and flats of chalcopyrite ($CuFeS_2$). The copper minerals leave characteristic green (malachite) staining which would have been visible on the exposed limestone. Copper mining could have been the reason for high status Iron Age activity nearby at Stanwick-Scotch Corner and there is also evidence of Roman metal working nearby, so it is possible they mined the ore here. The lack of factual data about the geological provenance of artefacts, metals or otherwise, is a generic issue and one that is rarely pursued in detail. Unless archaeological and geoscientific research works towards building a baseline of data and interpretation, then scientific and sociological questions related to the origin of the objects that fill so many display cases will remain unanswered.

Copper alloy oil lamp found at Arbeia (South Shields) fort.
Great North Museum collection.

The old shafts and diggings are either side of the road that leads to the church [NZ234057]

One of many mid-18th century shafts or diggings for copper on glebe land beside the church at Middleton Tyas

Benwell

The northeast of England is synonymous with coal. But did the Romans work coal here and if so where and what for?

Coal is dead plant matter that was turned into a rock by millions of years of heat and pressure. Around 310 million years ago, in the Carboniferous period, the environment of this piece of the north of England was tropical swampland, lazy rivers and lagoons. The plants in the swamps died and the resulting peat layers were buried under sand and mud, ultimately becoming coal seams. Beneath the Roman fort at Benwell (Condercum) are some of the 20-30 seams that form the world-renowned Northumberland and Durham coalfield. At the surface, on the valley side, 300 metres south of the fort is the thickest of those coals, the two metre 'High Main' seam.

Did the Romans work and use coal? Yes. There is evidence from many hundreds of places across Britain. Unfortunately, as the Roman word for coal, *carbo*, is also used for charcoal and wood, references in historical literature may be ambiguous and that can complicate things. Did the Romans actually mine it here at Benwell? A 2017 excavation within the civilian settlement (vicus) south of the fort located a building containing a large amount of coal. An earlier archaeological excavation had retrieved coal fragments which had fossil spore and chemical affinities with a seam which occurs beneath the site. Antiquarians in the 16th to 19th centuries believed that shafts and bell pits they excavated on the site were Roman. But others, including mining specialists, regarded the historical evidence as inexpert and equivocal and attributed the mine workings to medieval or later times, suggesting the Romans merely worked surface exposures of coal and not much of it at that. As with other minerals, archaeological evidence of extraction is often speculative, because any subsequent workings will have removed all trace of Roman ones.

From the evidence we have, the Roman military used coal all along the Wall. It is probable that it was also being transported some distance – to the Thames and the Fens – which would imply its widespread use and a very organised distribution operation. The assumption is that the major use of coal was for warmth, cooking and metallurgy, but there is evidence of its likely use in drying grain, manufacture of ceramics and brine evaporation, and even for cremation and jewellery. Is it possible other fuel sources – wood and charcoal – were unable to cope with increasing energy demands, especially in distant places whose populations were growing? Regrettably, coal, one of a few minerals we can trace the provenance of well, is often disregarded, mis-identified or not properly recorded in archaeological excavations. If, in future, digs record the detailed context of coal and submit it for analysis then we stand to learn more about the use and origin of a key Roman energy source and the socio-political stories it can tell.

Coal storage building beside Jenison Avenue in the vicus of Condercum excavated in 2017.
Image courtesy of The Archaeological Practice Ltd.

There are the remains of a temple and gated causeway across the Vallum within the residential estate [NZ217647]

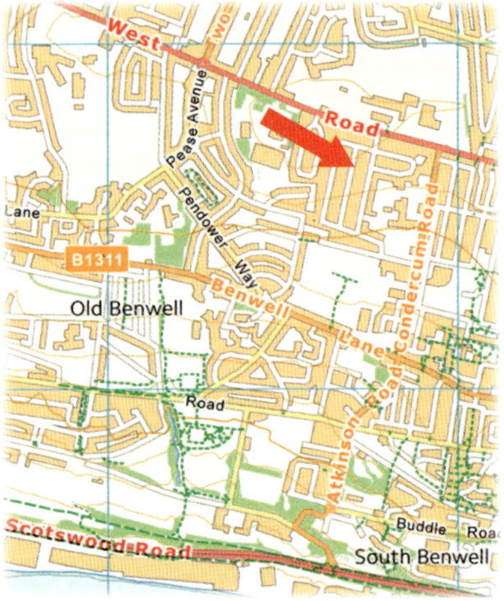

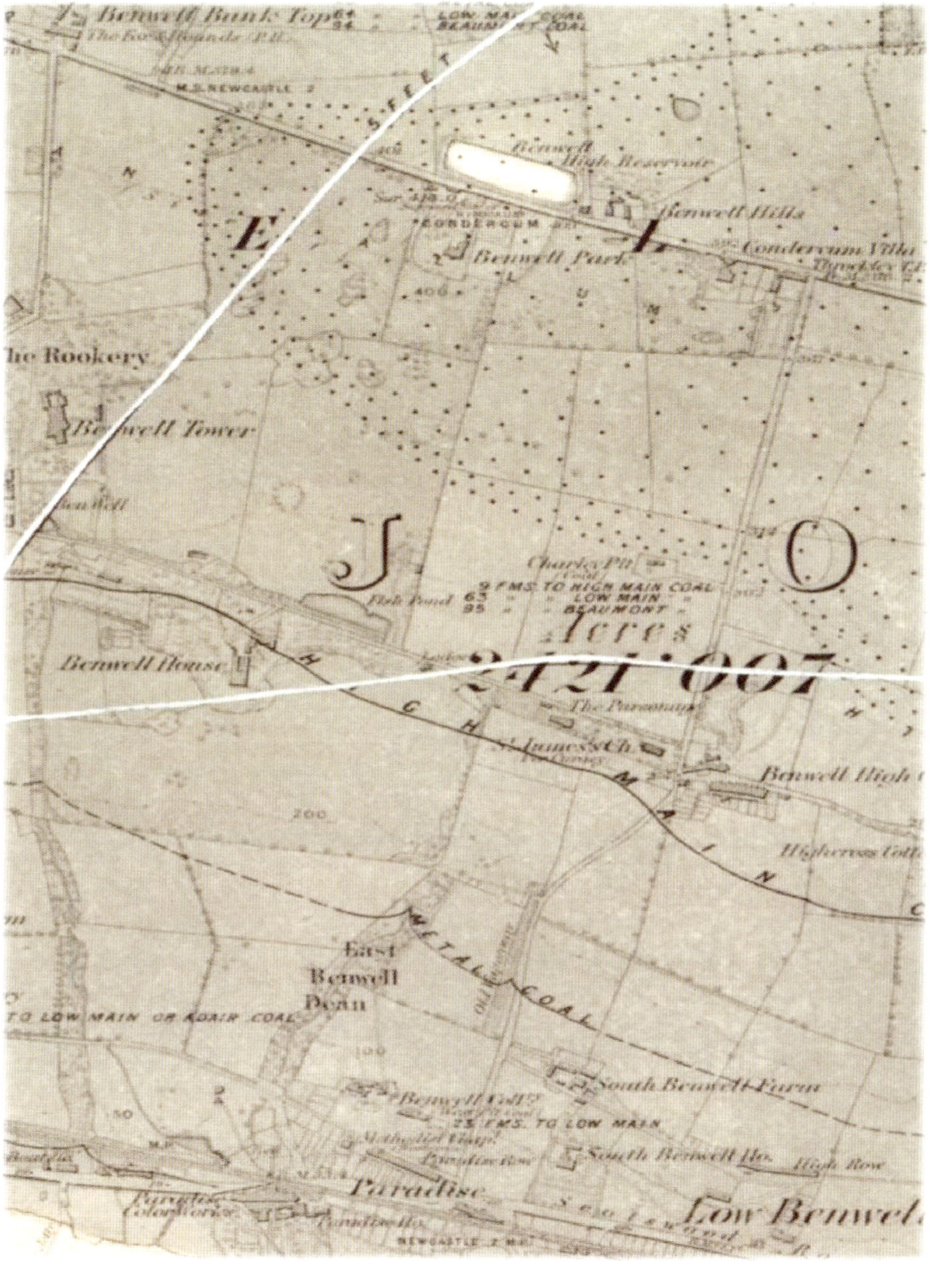

1:10.560 (6 inches to one mile) scale geological map of the Benwell area surveyed in 1866. The thick High Main coal seam is certain - a continuous black line - as the geologist observed evidence of its mining and is confident of its presence and position.

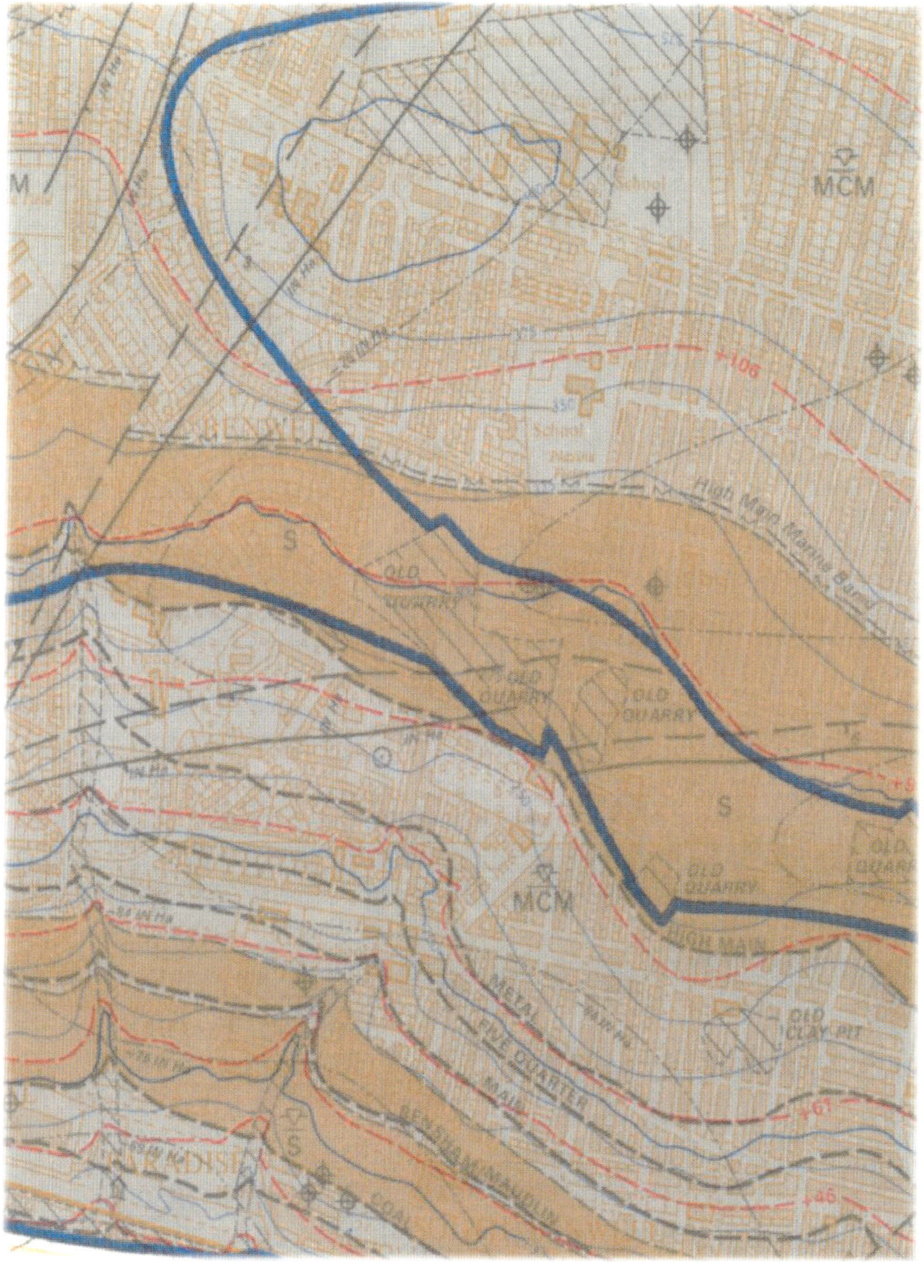

1:10.000 scale geological map of the same Benwell area surveyed in 1976. The thick High Main coal seam is uncertain - a dashed line - as old mining evidence is concealed by urban development and the geologist could not be confident of its position.

Vindolanda

Vindolanda is a perfect place to share another puzzle about the Romans and coal along the frontier.

The description of Benwell explains the geological origin of coal and how many workable seams underlie the Northumberland and Durham coalfield. But to the west there are other, older, Carboniferous coal seams. These may not be as numerous or as thick but they stretch from north of Alnwick to Brampton and Alston in Cumbria. Their outcrops straddle the lines of the Wall and the Stanegate for many kilometres. One of those older seams is exposed at the surface 500 metres southeast of Vindolanda and (like most coals) miners and geologists gave it a name – the 'Little Limestone Coal'. It was mined extensively in the 19th century at Barcombe Colliery. This same seam was also mined at Acomb, Fourstones, Bardon Mill, Henshaw, Haltwhistle and Blagill and is still mined today at Ayle near Alston. It is to this 'Little' coal, once a 330-million-year-old vast tropical swampy forest, that so many Tyne Valley and Pennine settlements owe their existence.

It seems inconceivable that the Romans were not aware of this coal seam. After all, they quarried the sandstone directly above it on Barcombe Hill and probably worked the adjacent limestone layers too. It is also possible that, unlike southern Britain, trees (and thus wood and charcoal) were not as abundant as a fuel here. Like many other Roman sites, throughout the years of excavations at Vindolanda pieces of coal have been found. It was likely used for heating, but also for metalworking and perhaps for lime burning. The intuitive conclusion drawn by many is that the coal used in Vindolanda was extracted from the nearby Little Limestone Coal seam.

However, the only piece of hard scientific evidence says otherwise. Analyses of fossil plant spores and a test of the petrological 'maturity' of a single Vindolanda coal sample points to it coming from west Durham, perhaps near Ebchester. Single samples of coals from Housesteads and Corbridge gave the same result. The scientist who did those tests was a very credible world expert on Carboniferous plant spores, but in each case they are single samples of coal. Even if he is correct about these samples, it is quite possible that quartermasters of forts along the Wall over 320 years of occupation sometimes ordered coal from Durham and sometimes sourced it locally. Another unsolved puzzle; but one we can unravel by gathering data on the archaeological context of coals, analysing their plant spores and maturity and building a database for correlation.

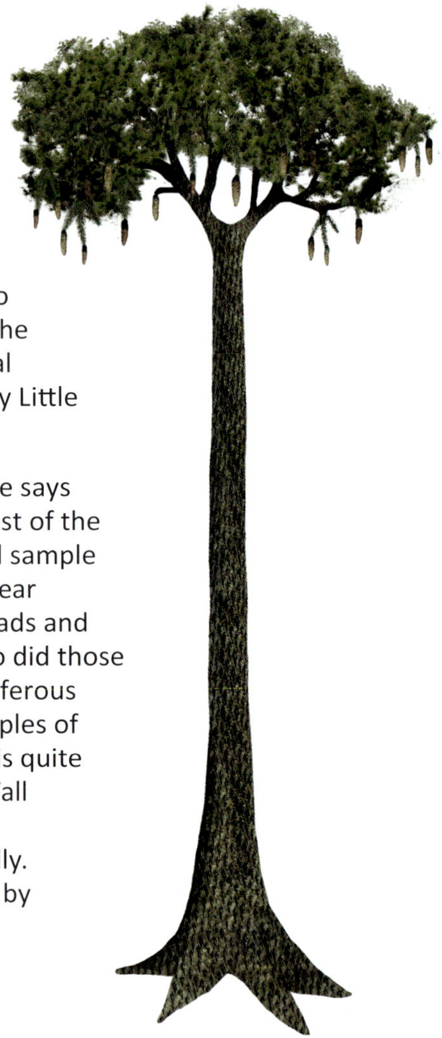

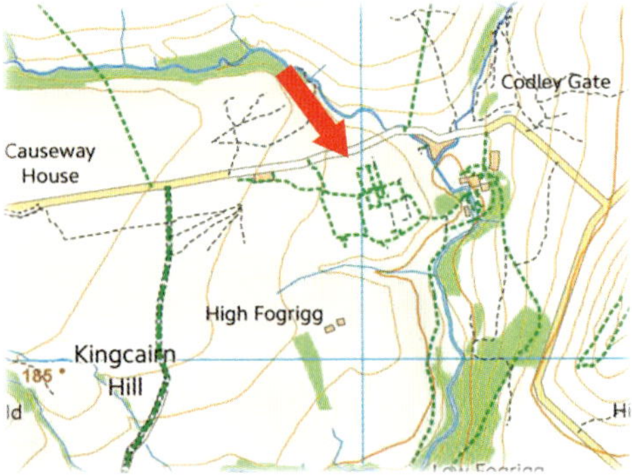

Vindolanda is well signposted from Bardon Mill and the B6318 and has ample parking. To visit the old mines take one of the footpaths south and east of the fort [NY770663].

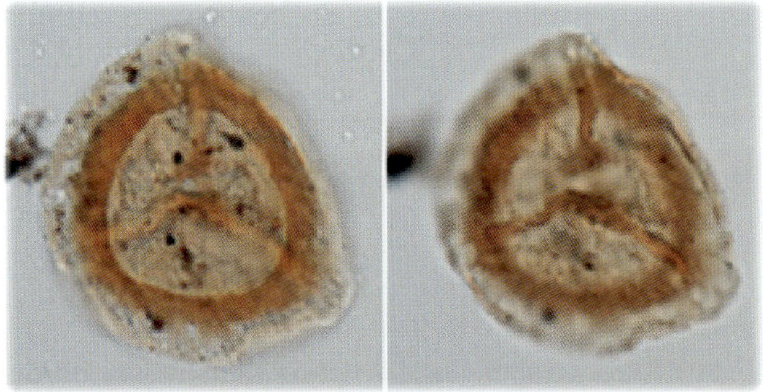

Fossil spores of Carboniferous coal swamp plant Lepidodendron and an artist's impression of the full plant. The spores are less than 50 microns (1/20th of a millimetre) across, the plant grew up to 50 metres tall.

Spore images by kind permission of Jan Hennisen. Plant image Tim Bertelink, CC BY-SA 4.0 https://creativecommons.org/licenses/by-sa/4.0>, via Wikimedia Commons.

Vindolanda, with waste from 19th century mining
of the Little Limestone Coal in the foreground

37

Peat – the forgotten fuel?

Swallow Crags

In the context of the Roman frontier peat is almost always thought of as a deposit which preserves things. Its use as a fuel is rarely considered.

Peat is partially decayed organic matter. It is most common in the upland and wetter areas of Britain and so extensive peat deposits – associated with mires, bogs and moors – are found across the north of England and southern Scotland. When dried out, peat makes a good fuel. It is still in widespread use in rural Ireland and not so long ago it was cut extensively for home fires in outlying parts of Britain too. 1500 metres north of the Wall and within sight of Winshield Crags is the sandstone escarpment of Swallow Crags. On its gentle southern slope a peat bog has developed, one of many either side of the Wall. Here you can pick out the edges of a typical shallow rectangular depression of an old domestic peat cutting. The tools and equipment used to harvest, store and transport peat can be found in nearby farms and in museums of rural life.

There is good evidence of the use of peat as a fuel to smelt lead in the Pennines in the 18th and 19th centuries where, turned into charcoal or mixed with coal dust, it produced a hotter flame. There are also indications that it was used to smelt iron, burn limestone to make lime for agriculture and mortar and evaporate brine in salterns. Looking back further archaeologists believe that early Bronze Age people in the Hebrides were using peat as a fuel and it seems probable that Iron Age societies and learning from them, the Roman army, made use of a readily available natural energy resource too.

The Roman chronicler Pliny described the cutting of 'soil' for fuel by communities who lived on the northwestern fringes of the Empire and peat blocks were found recently in a well in York backfilled in the 4th century. But firm evidence of Roman use of peat as a fuel is scarce. Not surprising given its often remote (and soggy) location; the fact that it is a very delicate substance; and that new peat accumulates at a millimetre a year as plants continue to grow. Elsewhere in the Empire the pragmatic Romans used animal dung and maybe even bone as fuel. In our wetter climate, with an abundant heat source on their doorstep, is it unreasonable to assume they used peat to provide warmth and fuel industry? Could it be, misquoting T. S. Eliot, that the lack of published evidence is not because it does not exist, but because it is challenging to find and has not been looked for?

The Little family cutting peat in one of the Cheviot valleys in the 1930's.
Photograph by kind permission of the Valley in the North Archive.

Park at Gibbs Hill Farm permissive parking place and walk west for 400 metres [NY743690]

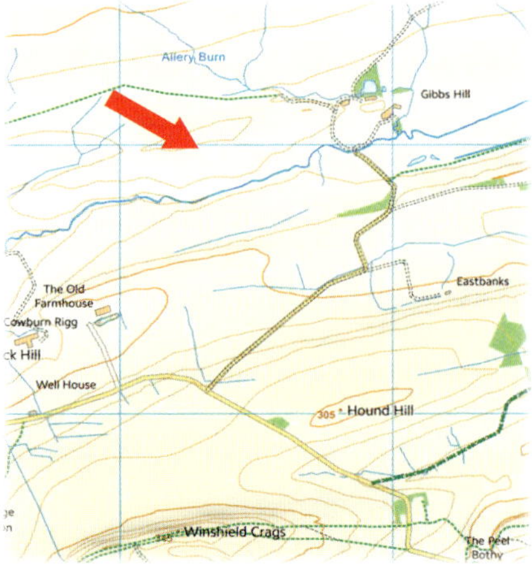

Peat cutting from around the 1940's at Swallow Crags, with Winshields Crag on the horizon

38

Crosscanonby

Sodium in small doses is an essential requirement of the human body. We get most of our sodium from salt.

Salt (sodium chloride – NaCl) was one of the first and most widespread traded commodities. The Latin word for salt (sal) is the derivation of the word for a Roman soldier's pay and ours: salary. This is also the origin of the idiom 'worth your salt'. Salt is thus a fundamental mineral and possession, but we don't know a lot about Roman production and trade of salt along the northern frontier. A report published in 2017 summarises evidence of Iron Age salt production and exchange in the region, which stopped abruptly at the end of the first century AD. The report states 'Paradoxically our knowledge of where and how salt was later procured in the Roman era remains very poor.' Salt comes from one of two sources: the sea, or rock salt. The latter is just a multi-million-year-old lithified version of the former, a product of evaporation in warm seas when Britain was much nearer the Equator.

In Europe the first evidence of salt production appears around 8000 years ago when people in what is now Romania boiled sea water. In the northeast of England, excavations near Berwick and Loftus have recorded Iron Age (and at Loftus, possibly Neolithic) salterns – seawater evaporation pans. Salt was much more than a condiment, it meant food, such as the fish and pork recorded in the Vindolanda tablets, could be preserved and transported long distances. Iron Age people and Romans exploited brine springs in Cheshire, at places we know by their salt-related 'wich' suffices, like Nantwich and Middlewich. The brine solution originated as dissolved rock salt layers deep below ground and was evaporated at the surface in ceramic and then lead pans. We also know of Roman sea water salterns along the south and east coasts. The assumption is that salt was a commodity that was transported to the northern frontier, sometimes in coarse pottery (briquetage) but there is scant evidence of that. There are many suitable areas for salterns along our northern coastlines and also a ready supply of fuels. There is a plethora of 'younger' salterns around the Solway Firth, some associated with 13th century monastic settlements, but if the Romans did make salt here in the north there are no signs..... yet.

On the Cumbrian coast, near Crosscanonby, a large post-medieval saltern is still clearly visible. Ironically, it is very close to Swarthy Hill Roman milefortlet 21. Two kilometres further south are rectangular rock basins and channels cut into the Triassic sandstones of the foreshore possibly related to a similar salt-making enterprise. The methods used in these places – collect seawater, heat and evaporate it, concentrate the brine and scrape out and gather the salt – are ancient and ones the Romans would have recognised.

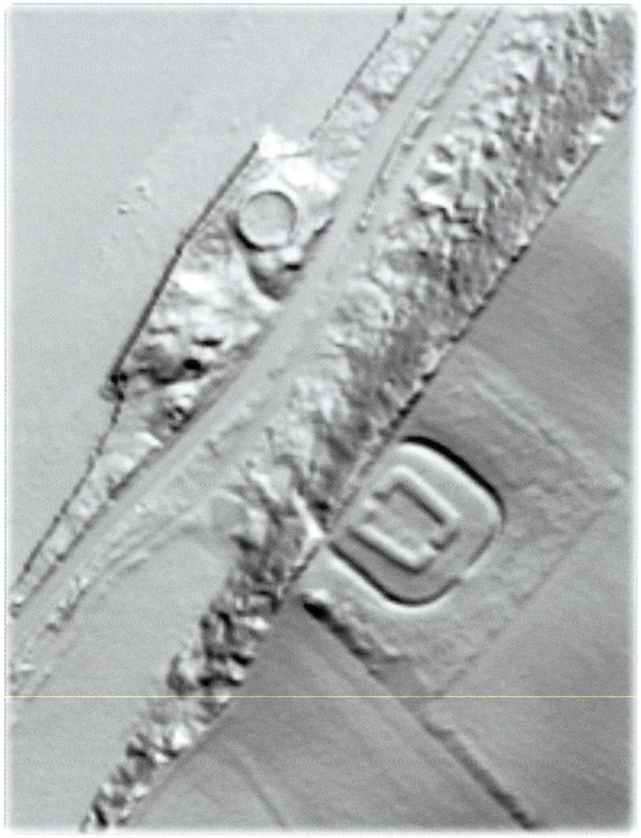

Milefortlet 21 and the saltern.
Digital surface model at >1 metre vertical accuracy.
Environment Agency data processed by GIS software.

The saltern and milefortlet are on the coast road north of the junction for Crosscanonby. There is a small parking area near the junction [NY671401].

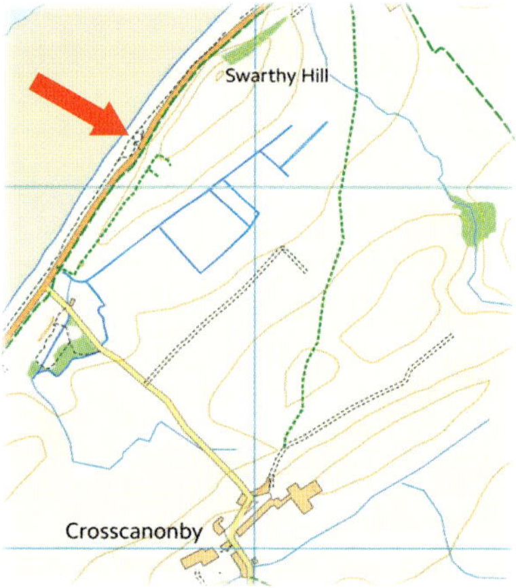

The saltern at Crosscanonby, with the Solway Firth and Criffel in the background

Chesters

Grain was a vital part of the diet of a Roman soldier. The large granaries in forts like Arbeia and Corbridge are clear evidence of that.

To process grain into food the soldiers needed rock – millstones or querns. The earliest types were saddle-shaped, one stone rocked over another to crush the grain. More efficient were rotary querns, initially just two stone discs with a handle, which became more sophisticated and diverse. As well as hand tools the Roman army also built mills driven by water and animals. Watermills have rarely been found along the frontier but one was discovered at Haltwhistle Burn near Cawfields quarry, its wheel race cut in Whin Sill bedrock. Water channels at Willowford Bridge and Chesters bridge may be evidence of mills there too. You can understand the need to mechanise; grinding by hand would take one man half a day to produce enough flour to feed the eight men in his barrack room.

Sandstone hard enough to make millstones is common throughout the local Carboniferous and Permian bedrock and many querns found at Roman sites along the Wall are made of it. The particle-size of the sandstones used varies and seems to matter less than their durability. Pinpointing where the sandstone to make querns comes from is as challenging as tracing Hadrian's Wall stone – sandstone is such a ubiquitous rock and there can be as much variation in composition in one quarry as there is across kilometres of outcrop. Sandstones with sand particles that are well cemented together made good millstones. Volcanic rocks, lavas with lots of small gas bubbles (vesicles) were even better – they self-sharpened and left fewer stone particles in the flour. Querns made of lava have often been found along the frontier and while similar rocks exist in Britain, these millstones are foreign imports.

They were quarried from very young (less than one-million-year-old) volcanic rocks at Mayen in the Eifel region of Germany and can be traced because their geochemical and petrological signature is relatively unique. The superior qualities of these querns meant they were a valued, traded item before Roman times and they were exported across the Empire. The legions prized Mayen querns and carried them on their campaigns. Mayen querns dominate Roman finds in Tongeren in Belgium, so it is no surprise that they were used by the Tungrian cohorts posted to Britain. Both lava and sandstone quernstones turn up frequently in excavations of Roman sites. They are often found in foundations and wells, sometimes deliberately broken – a ritual act that predates the Romans. Chesters Fort Museum at Chollerford is a great place to see the whole range, from the geologically youthful German imports to our more ancient local stones.

Detail of Mayen volcanic quernstone – gas bubbles (vesicles) are visible

Chesters Fort and Museum is about one kilometre west of Chollerford Bridge [NY912704]

Quernstones at Chesters Museum.
Image taken by kind permission of English Heritage.

Hexham

Much of what we know about the Romans is from their prolific carvings in stone: sculpture, inscriptions and even graffiti.

While there are Roman literary sources, they are few and often second-hand. The writing tablets at Vindolanda have been a revelation but they remain a rare insight. Roman inscriptions and carvings in stone shed light on so many things. Centurial stones tell us who built sections of the Wall. Altars, statues and tombstones record the gods they worshipped, their leaders and the lives of their dead. Not least, their graffiti carved in bedrock reminds us that in many ways people in the past are the same as people today.

In Hexham Abbey there is a Roman tombstone. It is large, about 2.7 metres tall. It was found face down under the floor of the Abbey in 1881 and like the stones in the Saxon crypt was robbed from nearby Roman ruins, almost certainly those at Corbridge. It shows a soldier riding a horse over the figure of a naked man. The carving quality is typical of the frontier, the makers were craftsmen, not artists and their raw material was rough sandstone, not Carrara marble. The tombstone, its inscription and symbolism, tells us about the soldier and his times. His Roman name was Flavinus but it may not have been his birth name. He was a first century cavalryman, a standard-bearer in the Ala Petriana. This was a unit originating in Gaul, which would become one of the elite regiments in the Roman army. He wears a tunic and cloak, a helmet with a plume and a torque (metal ring) around his neck. His weapon is a sword. He had joined the army when he was 18 and was 25 when he died. The carving has a recurring theme – domination of a cowering enemy, an ancient Briton perhaps. The memorial is large, did Flavinus make regular contributions to a burial fund to pay for it?

Like virtually all the monuments along the frontier this one is carved from northern rock. After almost 2000 years of both reverence and neglect its surface has worn smooth. There are few macro-sedimentological clues but on the left face are vertical lines – bedding planes. This fits, such a tall stone would usually be quarried parallel to the horizontal depositional plane of the rock, so it is now perpendicular to its original orientation. The true colour and characteristics of the sandstone are hard to discern. Superficially it has a pale grey-brown patina and looks local – Carboniferous – but on the right-hand face very small chips reveal red sandstone. Could it have been carved from the Triassic or Permian rocks of Cumbria? A more detailed analysis of the rock's mineral particles would tell whether it was from younger Cumbrian stone, but the ubiquitous nature of these sandstones would, as ever, likely frustrate attempts to uniquely connect it to the precise place it was quarried.

The Saxon crypt – constructed with Roman stones

Hexham Abbey is in the centre of the town [NY935641]

Flavinus' tombstone in Hexham Abbey

Carlisle Cricket Club

During recent excavations of a Roman building at Carlisle Cricket Club over 60 beautiful carved gemstones were found in drains. They had fallen out of rings as their owners relaxed in the hot, damp atmosphere.

Many of the gemstones are jaspers (deep red and opaque) and cornelians (translucent orange and red). Geologists describe these and other microscopically crystalline silica (quartz) minerals together as chalcedony. They are intaglios, so-called because symbols (usually of gods and life) have been delicately engraved into the stone. Given their small size and the tools and simple magnification available to their makers, the intricacy of the designs is astounding. As well as being objects of art and status they were used as a personal seal when impressed into soft material. Intaglio gems are found on many Roman sites across Britain but because of the numbers and types of stones recently found in Carlisle it is thought that it housed an intaglio workshop. The bath house (or perhaps a 'thermae') is itself is also changing perceptions of this Roman city. At 50 metres square it is said to be the largest building on the Wall. Two large carved sandstone heads and imperial stamped clay tiles were also discovered, more geological facets of the Roman presence here.

For geologists the prime and prosaic question is about the detailed composition of these intaglio stones? This is quickly followed by: can that information help us establish where they came from? It sounds straightforward but it is not. To confirm the origin of the gems requires sophisticated microscopic and geochemical analyses, some of which are micro-destructive. When the chemical elements of the stones are known, they could theoretically, be compared with a database of other gem analyses to define the location (analogous to the way DNA tests work). But unlike DNA, a database for gemstones made of chalcedony does not yet exist. The best that can be done is to compare individual stones with potential source areas and to combine that with historical information about the Roman gemstone trade.

It is frustrating because it means we don't definitively know if these stones were imported (perhaps from as far away as Asia or Africa?), or whether some were sourced in Britain. Geologists have long known there are places near Roman sites where chalcedony occurs. Jasper and agate (a banded silicate) are found infilling small cavities (vesicles) in Devonian and Carboniferous lavas in the Cheviot hills, in Scotland in the Campsie Fells and at Birrenswark. After many years of weathering of the lavas chalcedonies like these are often found within the shingle of nearby streams and beaches. Until the scientific analysis can be done, we will have to be content with admiring the beauty and craftsmanship of these artefacts and reflect on the insight they give us into the beliefs and lives of people almost 2000 years ago. For many that is enough.

The excavation at Carlisle Cricket Club in 2023

The Cricket Club is northwest of Eden Bridges [NY400567]

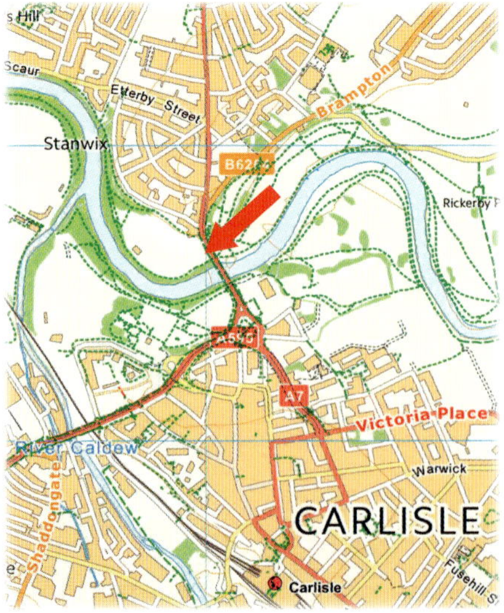

Intaglios found at Carlisle Cricket Club.
Image by Anna Giecco.

Great North Museum

The Romans were just as dependent on the Earth's geological resources as we are. If they could not grow it then they would mine or quarry it.

To appreciate the extent of that dependence, visit any museum with Roman artefacts. The display at the Great North Museum (Hancock) in Newcastle is exceptional. The objects range from the spectacular and beautiful to the obscure and utilitarian. Search out a gold ring found during an excavation at Corbridge Roman town in 1840, the workmanship is exquisite. It is thought to be a love token or betrothal gift made around 200 AD. The cut-out letters read 'Aemilia may you live'. Gold does occur in the rocks of northern England but it is so disseminated and in such miniscule quantities that it is only of academic interest. At Dolaucothi in southwest Wales there are workable amounts of gold and clear evidence that the Romans (and maybe even earlier societies) mined it extensively.

We know the Romans did work one northern rock for jewellery – jet. It is a black, dense form of fossilized wood which is found exclusively in Jurassic shales north of Whitby. It was highly prized long before Queen Victoria made it fashionable, perhaps even in Neolithic times. Jet is easily cut and is capable of a high polish. Thought also to have medicinal and magical qualities it was much valued by Romans and their historians, like Pliny, were fascinated by it. Jet jewellery has been found throughout Britain and in many sites along Hadrian's Wall. While it has been extensively mined, the Romans probably collected it from the beaches. Jet is a lot less plentiful there today, but Deepgrove Wyke foreshore is as good a place as any to try your luck. Please make sure the tides are right!

The Romans worked lithomarge (Greek – 'stone of marl') into beads, pendants and amulets and believed it too had magical and religious properties. Less mystically, lithomarge is a product of weathering of feldspar- and iron-rich rocks. To geologists it is kaolin, a rock composed of clay minerals: aluminium silicate and oxides and hydroxides of iron. It often has a pale white matrix marbled with brownish purple veins. The stones are thought to be continental, perhaps from the north coast of the Mediterranean, but weathered hematite ores of southwest Cumbria offer a potential local source.

Whetstones may seem mundane, but until a few decades ago they were essential everyday tools. Cheap replacement of knives and chisels was not an option and they needed sharpening. Whetstones were equally indispensable to maintain the Roman army's weapons and edge-tools and they are found in almost every excavation. All rock types appear to have been used: igneous, sedimentary and metamorphic, but detailed geological identifications in archaeological reports are rare and so whether whetstones were sourced from near or far is usually unknown.

Whetstone made of Carboniferous sandstone (South Shields)

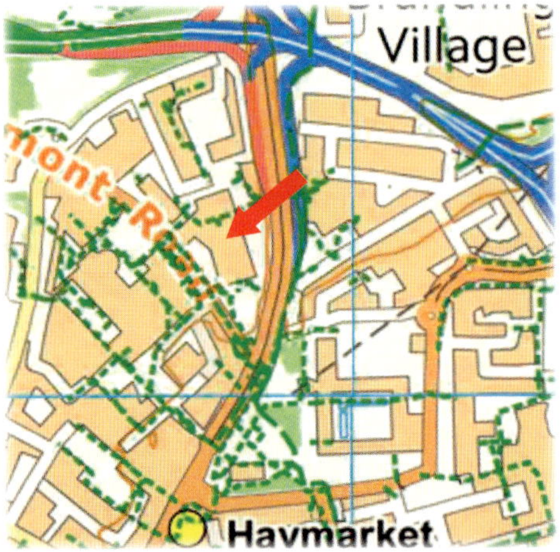

The Great North Museum is in the centre of Newcastle [NZ248652]

Beauty and utility

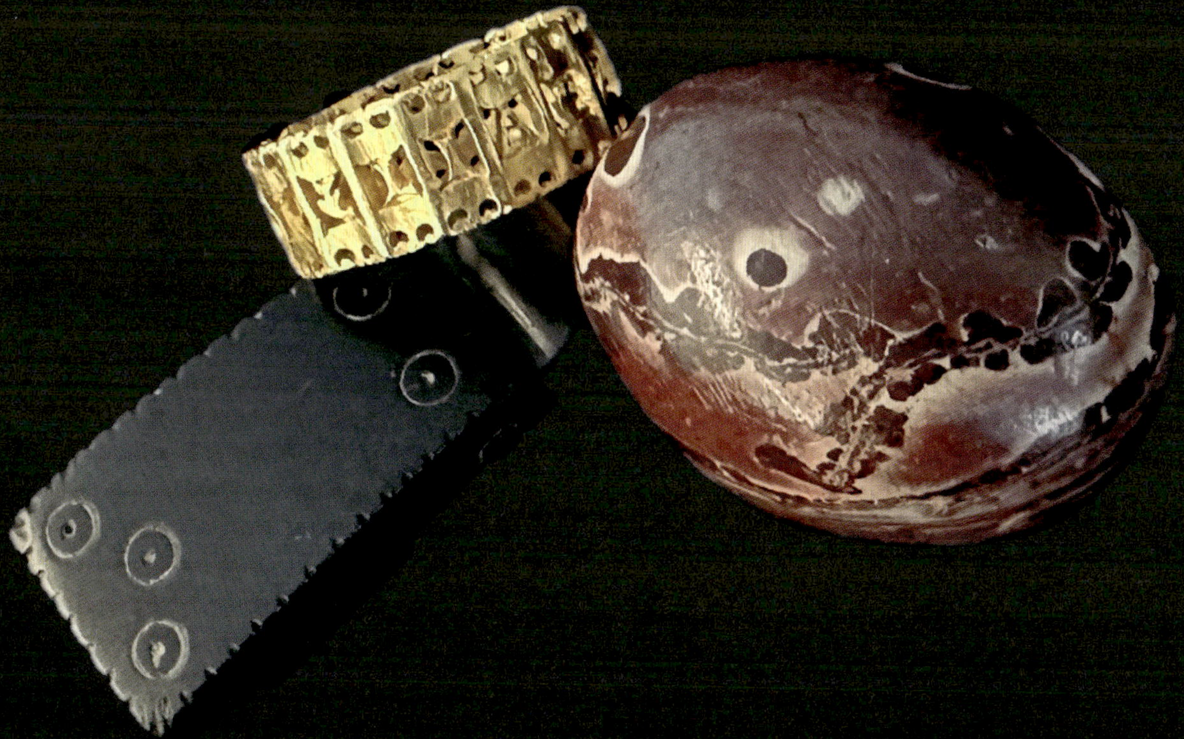

Aemelia gold ring (Corbridge), Lithomarge amulet (Housesteads), Jet pendant (South Shields).
All in Great North Museum, Newcastle upon Tyne.

Arbeia

The scale and grandeur of Hadrian's frontier is extraordinary but it is sometimes difficult to imagine it as it was 1900 years ago. Arbeia Fort in South Shields gives us an insight.

Even those of us who are passionate about the Wall monuments acknowledge that it isn't easy to translate the grey-brown stones of its skeleton into the living landscape it used to be. But these Roman buildings have not always been the truncated, muted remains you see today; many were covered in white lime render and painted in a variety of colours. While Arbeia has attracted controversy in archaeological circles, for most visitors the full-size reconstruction of buildings and interiors will bring the frontier to life. The size of the rooms and the height of the ceilings in the late Roman courtyard house are a surprise but it is the walls with their interpretation of fresco painted patterns and motifs that may be the biggest challenge to preconceptions of Roman life at the edge of the Empire. Vibrant colours are not something we associate with the Wall and its monuments and yet painted plaster fragments from Arbeia and villas in southern England show they were used.

The dull sandstone and wattle and daub walls were plastered with lime render made by burning local limestone in kilns and mixing the resulting quicklime with water. Several layers may have been applied and the internal surfaces of walls in high status buildings may have been built up with finer coats before finishing with whitewash. The painted patterns and motifs, frequently shades of red and orange-brown, were produced from natural pigments. Some were plant-based but others were extracted from rocks.

Ochre and umber are derived from iron minerals within iron-rich rocks like ironstones and red sandstones, or from mineral veins and their weathered residues. All these are common across our region and quantities of umber and ochre were still being mined near Alston around 140 years ago. Green and especially blue pigments, are rare and some may have origins in copper minerals like malachite and azurite, both of which are found in the north of England. White pigments were produced from limestone, gypsum and lead, all in abundance locally. Pigments were not just used on walls; they have been found colouring inscriptions on altars and were used for dyes, cosmetics and medicines. Geochemical analysis that might help distinguish between local and imported pigments is expensive and rarely done and so, like intaglios, telling stories about their origin and trade must wait.

Hard ironstone (limonite) on a mine dump at Horse Edge, southwest of Alston. When weathered and soft it forms the umber they mined hereabouts.

Arbeia (South Shields) Fort is reachable by Metro, bus and car [NZ365679]

Reconstruction of the commanding officer's courtyard house (and dining room) at Arbeia - South Shields

Lanchester

Just west of Lanchester is a Roman fort, Longovicium. Supplying the fort with water was a network of aqueducts and dams, testimony to the skills of Roman engineers and their understanding of the landscape.

To build such infrastructure today would require a multidisciplinary team. The design, plan, calculations and installations would need civil, geotechnical and mechanical engineers. Understanding the shape of the site, the rocks and water dynamics and where the dam materials would come from is the work of surveyors, hydrologists and geologists. The Romans had all these skills within their legions, but they were not working from scratch: a comprehensive range of specifications had been devised by engineers like Vitruvius in the first century BC. When the Reverend John Hodgson, curate of Lanchester and an antiquarian, saw them in 1805, the series of aqueducts and dams in the valleys west of the fort were "as visible as the day they were made". His investigations were published in Archaeologia Aeliana, the journal of the Society of the Antiquaries of Newcastle upon Tyne he helped to found. 130 years later, when another giant of archaeology, Kenneth Steer, explored the structures as part of his PhD they were evident but much less apparent. Subsequent damage by agriculture, forestry and coal extraction means that the aqueducts have all but disappeared but the dam can still be clearly seen.

The Roman engineering starts five kilometres west of the fort in the small valley of the Dykenook Burn. This valley, possibly a glacial meltwater channel, is cut into Carboniferous rocks: sandstones and shales containing two significant layers – the Brockwell Coal and the German Bands Ironstone. These two rock units also occur near Longovicium fort and were exploited extensively from the 16th century and perhaps long before. The dam site was surveyed in detail most recently in the 1980s by a local civil engineer. He recorded it as around six metres wide and 100 metres long. In plan it is a curve, a sideways arch with a radius of 45 degrees, convex upstream. The survey noted the remains of the stone-block facing and that the core was a clay base, with sand above; they speculated that it may be one of the earliest ever 'arched' dams. Open channels, aqueducts, connected this dam with another and then a series of aqueducts carried water to the fort. At one point one of the aqueducts must climb upslope. The hypothetical solution to this obstacle, suggested by Kenneth Steer, was a siphon system using lead pipework; the Romans had used these elsewhere.

Why did the Romans build such complex structures to supply so much water? It seems excessive for the usual domestic and bathing water needs of a fort. Could the reason be that, with local coal and iron ore available, water was the third essential ingredient needed to smelt iron? The Reverend Hodgson also recorded thousands of tons of slag in the area and numerous hearths, cinders and slaking troughs around the fort....

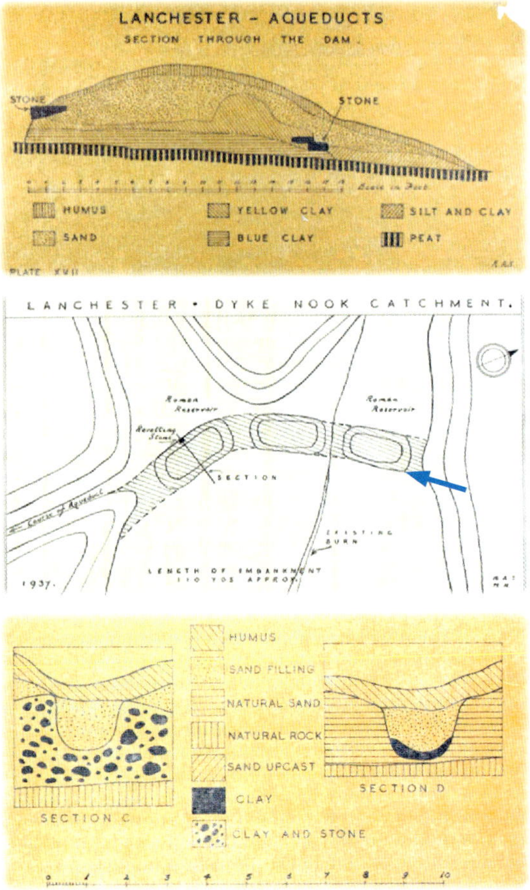

Plan and sections of the dam and aqueduct in the Dykenook Burn valley.
From Kenneth Steer's 1938 PhD thesis: The archaeology of Roman Durham http://etheses.dur.ac.uk/683/.

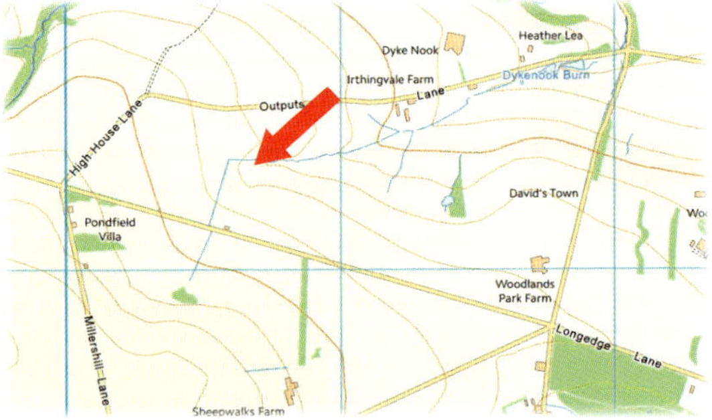

The dam is in the Dykenook Burn valley south of Outputs Lane [NZ107474]

The Roman dam in Dykenook Burn valley from the northeast
(viewpoint marked with blue arrow on diagram opposite)

Coventina's Well

Water is an essential human need but it also played a powerful role in the spiritual lives of Romans.

Water and its associated engineering get a lot of archaeological attention in drier parts of the Empire but less so in Britain because of our high annual rainfall. That is even more true here in the north where our wetter climate means that surface water resources – streams, rivers and lakes – are abundant. Many northern Roman sites have watercourses nearby and rainwater is also channelled from roofs into storage cisterns, but the extent to which Romans depended on groundwater and how much they revered it is often overlooked. Coventina's Well is just west of Carrawburgh Fort, known to the Romans as Brocolitia. An 1876 excavation revealed water feeding a rectangular basin in the centre of a walled enclosure or temple. The 'Well' (more a spring) emerges on the northern limit of an outcrop of the Whin Sill dolerite, and as this rock is impermeable it is probable that is why groundwater comes to surface here. Almost 14,000 coins and many stone altars and inscribed slabs were recovered from Coventina's Well.

At Housesteads a spring emerges naturally at the top of a sloping bed of limestone 200 metres south of the fort. It flows into a stone-lined tank within a structure believed to be a shrine to Mars Thincsus and the Alaisiagae, German war deities worshipped by Friesian soldiers. There are wells recorded at many other sites, for example at Benwell (Condercum) and at Vindolanda and Magna. Votive objects and inscriptions are not the only things that have been found in Roman water sources. Prior to the Lanes redevelopment in Carlisle around 1980, excavation of a stone-lined third century well exposed the skeleton of a man who was probably murdered and then disposed of. You can read this story and even see his facial reconstruction at Tullie House Museum. The same excavation also revealed a second century barrel-lined well and waterwheel and in the 1920s at the Blue Bell Inn in Scotch Street, another well more than 12 metres deep was found. Based on recent boreholes these wells are likely to have struck water at about 5.5 metres down in glacial clay and stones, before hitting 250-million-year-old Triassic mudstones at around 7 metres.

Most often it is the artefacts that infill the wells that have been studied, but the basics of Roman water and groundwater supply, use and drainage also deserve scrutiny. This includes everything, from how they prospected, dug and constructed their wells, to how they dealt with the challenges of ventilation, stability and contamination. Then we would understand better what Romans knew of water sources, how they supplied enough of it to satisfy the essential physical needs of the thousands of occupants of their frontier settlements and maybe, more about their spiritual desires too.

Site of Coventina's Well

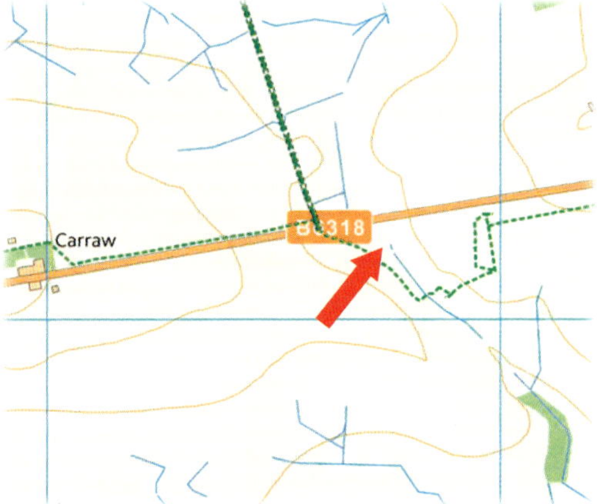

The Well is immediately west of Carrawburgh Fort and Mithraic Temple [NY857712]

Sandstone relief carving of three water nymphs found at Coventina's Well. Now in Chesters Museum.
Image taken by kind permission of English Heritage.

Bewcastle

Bewcastle village and its Roman fort, remote outposts in northern Cumberland, tell the story of the use and re-use of stones that connect us with our past.

Bewcastle today is a small, quiet, dispersed settlement in the valley of the Kirk Beck. It is centred on a small flat-topped hill of sand – outwash from the melting of the last ice sheet. Surrounding it to the north, east and south are high, rough fell-land of repeating horizontal layers of sandstone, limestone and shale dating from the oldest part of the Carboniferous period 350 million years ago. The well-drained sand hill in the valley must have been an attractive site for the earliest prehistoric settlers and it is said there was a shrine to Cocidius, a local god, here. Around AD 122 a Roman legion built an earth and wood fort at Bewcastle at the end of the Maiden Way, which connects it to Birdoswald. Stone signal stations on prominent sandstone scarps at Barron's Pike and Gillalees Beacon provided vantage points for the fort and the road.

At least two auxiliary cohorts were garrisoned at Bewcastle, the Dacians (from Romania, Bulgaria and Serbia) and the Nervii (from Belgium). From artefacts found on the site they adopted Cocidius as their god too and the Roman name of the fort links to that: Fanum Cocidii. Over the next two centuries the fort would be rebuilt in stone, its walls defining an almost hexagonal perimeter mirroring the plateau. Possible sources of building stone are quarries at Crossgreens to the north, one in the aptly named Earthwork Sandstone. The fort had a turbulent history and was destroyed and rebuilt several times, finally being abandoned in the fourth century.

In the centuries that followed, Bewcastle continued to draw people. It became an Anglo-Saxon settlement and in the churchyard is a richly carved stone cross shaft dating from the late 7th or early 8th century. The cross, one of the finest to survive, may well have been made from stone robbed from the Roman fort but there are also primary sources for such a monument in the thick, older, sandstones that surround the village. The 13th century church, St Cuthberts, has much re-purposed Roman stone too. The Vikings were here and the modern name of the village may derive from Beuth, a Norse chieftain. Sometime after the 1092 annexation of Cumberland from Scotland, 'Bew Castle' was built in the northeast corner of the Roman fort, largely from its stones. The castle's position on the front line meant it was a busy place during the wars between the Scots and the English and it was also rebuilt multiple times. It was garrisoned for the last time in 1639 to deal with 'commotions in Scotland' and dismantled by Parliamentary forces in 1641. Visit Bewcastle, you may well have all this history and the stones that tell it, to yourself.

The ruins of the 11th century 'Bew Castle' constructed of stones from the Roman fort

Park in front of the church, which is within the Roman fort. The cross is in the churchyard and the castle is a short walk away [NY560747].

Bewcastle and its forts from Tower Brae

Catterick

47

If we can identify the constituents of the rock that an object is made of it may be possible to identify where it came from – its provenance.

As bone and teeth are made of minerals this applies to them too. Tracing people and their things is crucial in understanding ancient civilisations. In the past attempting to define provenance of materials like marble, or artefacts such as pottery, relied on visual inspection of their colour, density, texture and hardness. Optical microscopes give us even more information about the size, shape and composition of the particles and minerals in rocks. In coal, peat and sediments we can identify the types of microscopic fossils: spores, pollen and diatoms. In the last 30 years more sophisticated techniques, like scanning electron microscopy and X-ray fluorescence, have allowed determination of elements down to a few parts per million. Now it is routine to use isotopic ratios to precisely measure the molecular and atomic make-up of artefacts to trace their origin, whether they are raw materials, finished products, or human tissue. Because our bodies absorb traces of the rocks around us in our food and water, the geochemical signatures they leave in our bones and teeth provide clues about how and where we have lived.

The results can be revelatory but these techniques are not infallible. Often, because many rocks are not unique, it is extremely difficult to definitively identify the source. The same is true of metals and their ores, further complicated by their processing and reuse. There are other challenges: the need to establish where all relevant geological sources are; sufficient understanding of the stratigraphic and historical context of the artefacts; doing enough analyses to ensure the results are statistically valid; the lack of comparative national and international databases; the destructive nature of some tests; and not least funding the analyses.

Roman burial sites found recently at Bainesse, near Catterick, are an example of the capabilities of the use of multiple isotopes, but also the complexities of such analysis. Teeth and bone samples from many skeletons were analysed for a variety of isotopes. Strontium isotope variations are related to the bedrock in the area where the food eaten was grown. Oxygen isotopes reflect the water we drink. Carbon and nitrogen isotope variations in our skeletons relate to the proteins we consume. The analyses showed that compared to major cities like York, which have more diverse populations, the majority of individuals found in Catterick were local. But at times the evidence from isotopic analysis and the cultural artefacts and burial rites diverged. The research has underlined the importance of context and the need to combine multiple strands of evidence if we want to untangle human origins. Geochemical fingerprinting can be invaluable, but it is not a universal silver bullet.

The Roman cemetery at Bainesse, Catterick.
Image © Northern Archaeological Associates.

If you have used the A1 since 1959 you have driven through Roman Catterick - Caractonium [SE229991]

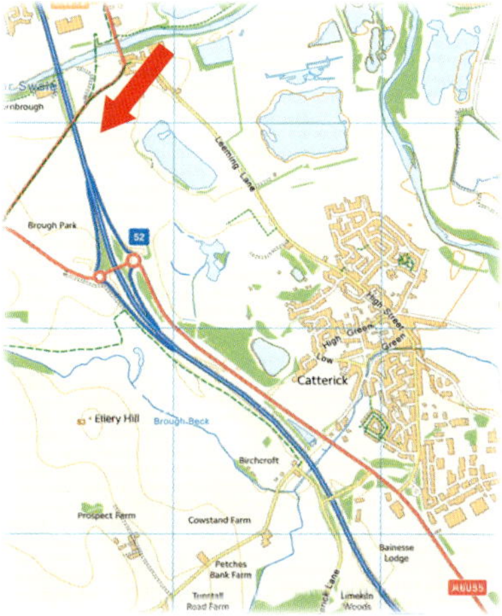

Where

Lower mandible from Coldrum in Kent. The individual is dated to 464–603 AD. Teeth were analysed using multi-isotope techniques to evaluate the provenance of the remains.
Image courtesy of Samantha Neil, University of Oxford.

Steel Rigg

48

For many visitors their first encounter with Hadrian's Wall begins at Steel Rigg.

That first awe-inspiring view east of the Whin Sill and the Wall takes everyone's breath away but a deceptively simple question quickly follows: when was the Wall built? Rough answers are easy enough: around AD 122. Precise and accurate ones are a lot harder. Dating things can be a challenge in archaeology and geology and generates a lot of debate and speculation. Both disciplines often use the same tools to try to answer when. Dating can be 'absolute' or 'relative'. Absolute dating tells you how old something is against a specified archaeological or geological time scale. A common absolute method in archaeology (but not often in geology) is using literature and producing historical chronologies. Other, scientific, dating techniques are used by both domains and include radiometric (e.g. Carbon -14), thermoluminescence, fission track, electron spin resonance, palaeomagnetism and dendrochronology. These tools each work best on particular materials. Relative dating on the other hand tells you whether something is older or younger but not exactly how old it is. Stratigraphy is probably the most well-known, it is based on the rule of 'superposition' – the youngest material is on top or cuts across others. Other relative methods are based on understanding a changing sequence, which could be the evolution of fossils, differences in style or diagnostic artefacts, and the classification of distinctive spores and pollen grains.

In order to date the Whin Sill dolerite both absolute and relative techniques have been used. It occurs as a layer within Carboniferous strata but is not present within the younger Permian rocks. Early geologists thought it was lava that had flowed across the ground surface, but metamorphosed sedimentary rocks directly above it proved the dolerite was an intrusion into them and thus later – younger than 300 million years. In recent times dating using radiometric and palaeomagnetic techniques has shown it to be between 294 and 300 million years old....and that there is more than one Sill!

But back to when was the Wall built? The most precise dates come from scrutinising literature and inscriptions and interpreting the clues they provide. Needless to say, there is more than one interpretation. Hadrian is thought to have visited Britain in AD 122 and some see this as the time when the Wall was decreed and begun. Others argue that the Wall was ordered from a distance and begun in AD 120 or 121, with Hadrian inspecting the work in progress in AD 122 and ordering various changes of plan. The evidence of coins, inscriptions and ancient texts can be used to support either explanation. It is a topic on which rocks offer only indirect and imprecise evidence, after all, we geologists are usually happy with estimates that are plus or minus a few million years.

Whin Sill dolerite on the east side of Peel Gap. The holes were drilled by researchers taking samples for palaeomagnetic/radiometric analysis (who regrettably failed to follow conservation guidance and refill them).

Use the National Park car park at Steel Rigg and walk southeast for 150 metres [NY75367]

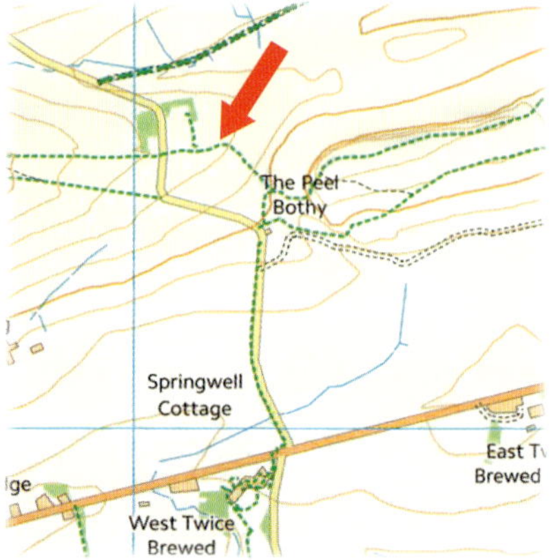

The view east from Steel Rigg: Peel Crags with Crag Lough in the distance

Magna

49

The discovery and preservation of archaeological remains often seems to be serendipity, but we now understand much more about the reasons why they survive in some places but not others.

Artefacts found on land are usually in ground that humans have disturbed, whether in life or in death. Sometimes that zone extends down into 'natural' geology – Quaternary deposits, or even bedrock. But mostly objects stay within relatively shallow surface layers, the soil and subsoil. These surface layers commonly inherit many of their characteristics from the geology beneath, at its most basic whether they are acid or alkali. If you look at a soil map you will see it mimics a geology map. Other factors do come into play, for example, the water table and subsequent human interference. How well different archaeological remains survive depends on the properties of the material they are found in, the way their physical and chemical composition reacts to their place of burial and how land-use changes like agriculture, development and changes to the groundwater system impact the area. Acid soils preserve pollen well, but are detrimental to metals, bone and teeth, which all survive longer in alkali conditions. Everything is better preserved when oxygen is excluded – i.e. in (anaerobic) waterlogged ground. This environment conserves deposits like leather, clothing, wood, and plants. Such sites are rare in many parts of Britain but the north of England is blessed. The acidic and wetter conditions found in places here mean that while bone, shells and ferrous and copper metals are often poorly preserved, invaluable organic objects survive much better.

The fort at Magna (also known as Carvoran) is an excellent example of archaeological preservation and its challenges. Thought to have been established as a Stanegate fort around AD 80, it was developed into a fort on Hadrian's Wall and was occupied until the end of Roman Britain. It is on the junction of two Roman roads and recent geophysical surveys have indicated extensive settlements around it. There may have been as many as 500 soldiers, at different times from Syria and Dalmatia and many more supporters and civilians. Some preliminary investigations 50 years ago found timber remains under a stone gateway. Twenty years later within deep anaerobic conditions Roman leather was encountered. Between the fort and the Wall today is a peat bog. It (or a lake that was there before it) was seemingly large, deep and wet enough to persuade the diggers of the Vallum to divert their ditch.

The peat bog is no longer obvious but it was to geologists in 1879 and they included it in the first map of the area. They also delineated glacial deposits and the Whin Sill and limestone layers beneath those. Such geological factors are critical to any hydro-environmental assessment of an archaeological site. This bog is now drying and shrinking and the topography around the fort changing. Magna's 17 hectare scheduled monument is considered likely to contain rare anaerobic conditions comparable to those found in Vindolanda and Carlisle; places which produced stunning discoveries. Areas of organic preservation at Magna are believed to be at serious risk and a five-year programme of excavation and research on the changing environment started in 2021.

A lady's studded leather shoe sole from excavations at Vindolanda, demonstrating how well anaerobic conditions are able to preserve organic objects.

Magna/Carvoran is 1.5 kilometres northeast of Greenhead [NY667659]

Digital scan of County Series geological map surveyed around 1879 (original scale: 6 inches to 1 mile).
Inset photograph looks northwest over peat bog and environmental monitoring station. Blue arrow = view point.

50

Raeburn Flow

Raeburn Flow is a small Scottish peat bog 12 kilometres north of Hadrian's Wall. There are many bogs in the frontier zone but this one has clues that the Romans mined lead.

The Flow rests on glacial stony clay which in turn overlies Triassic St Bees Sandstone. It is an ombrotrophic bog, meaning the plants and mosses in it receive all their water and nutrients from rain and not springs or streams. This means the plants are isolated from the surrounding landscape and are ideal for detecting elements in the atmosphere – including pollutants. Taking core samples from the peat in these bogs, analysing them botanically and geochemically and then correlating the results with radiocarbon dates of the peat, allows the detection of any human-induced contamination (eg ash, dust, metals) and its calibration with time. A core 364 centimetres deep was analysed at Raeburn, it represents approximately 3300 years of peat accumulation, around 1 millimetre a year.

The results showed a peak in lead pollution starting in the late Iron Age, expanding during the Roman occupation, and then declining. Exactly where the lead was coming from is unknown. The North Pennines are a strong candidate, but earlier sampling of peat in Teesdale and Weardale did not detect a similar lead signal. There are lead ores in the Lake District and southwest Scotland that could also have been mined and smelted. These methods cannot distinguish between metal mining and metal working but the research here is another persuasive clue to help solve the mystery of the source of lead on the northern frontier. To date the number of bogs and lake sediments along the frontier that have been investigated by these methods are few and there are many that have the potential to contribute to our understanding of mining, vegetation and climate change.

The methods used at Raeburn have, however, been used in other peat deposits and lake sediments across Britain and Europe. They show a widespread, consistent pattern: metal contamination in the atmosphere goes back to the Bronze Age, rises in the Iron Age and peaks in the Roman era, before declining and then growing significantly through medieval and later times. The signals of metal and other pollutants in our atmosphere are so dependable that they can be used as a proxy to date other events – a technique appropriately named 'wiggle matching'. There is a more fundamental application of these techniques – climate change. Deeper cores are being taken in locations that can take us much further back in human history. Crawford Lake in Ontario, Canada, has a virtually complete record of humanity's impact on Earth, from forest clearance, to metal smelting and fossil fuel burning and nuclear power. So profound are the changes we have made to our planet in the last few centuries and decades that many geologists believe, despite its incredibly short span, that we should define a new period of geological time – the Anthropocene.

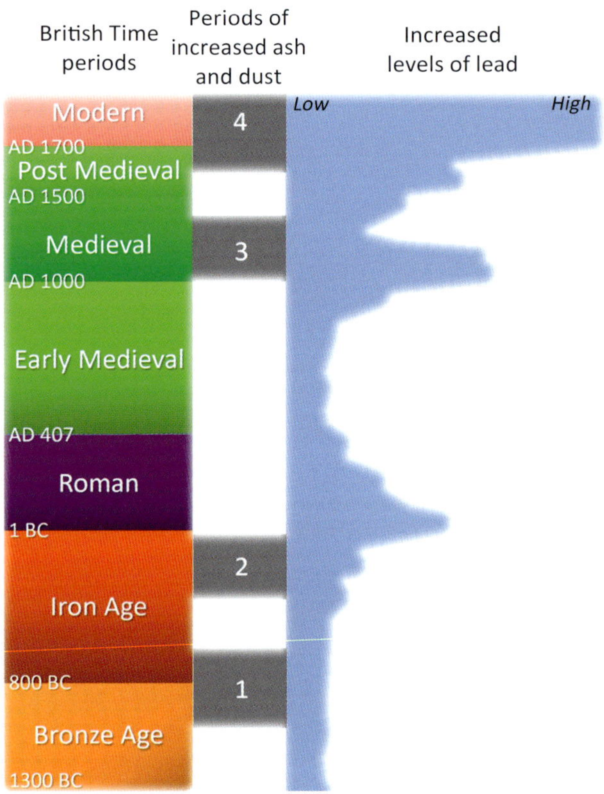

The variation of ash, dust and lead levels in the atmosphere recorded in Raeburn Flow.
Diagram extract by kind permission Tim Mighall.

The Flow is adjacent to the B6357 road northeast of Kirkpatrick Fleming [NY294718]

Raeburn Flow and scrub land at its edge

Endnote

I was raised at one end of Hadrian's Wall and went to university at the other. The Wall has been a persistent thread in my life and even when living well to the south I was a remote sponge, absorbing anything about this northern landscape and particularly connections between its rocks and the Romans.

There was little doubt that on retiring to live near the Wall in 2012 that my interest would grow and in the last few years my passion for things at the interface of geology, archaeology and history has deepened and widened. But I never anticipated that curiosity going into print; it was the recent success of the county Rock books that changed that.

There are two ways this book could have been produced. The first would have been to research each of the topics in great detail, publish in academic journals and then write a synthesis. This would have undoubtedly delivered a more thorough and considered work but the breadth of the field, the dearth of prior research in this cross-over domain and lack of resources and time (my age!) meant this was not feasible. The second path and the one chosen, was to produce a broad synopsis of the field, one that was as scientifically accurate and up to date as possible.

Many questions must be left unanswered – frustratingly, despite decades of dedicated archaeological investigation, insufficient scientific data and evidence exists to establish the origins of so many materials..... and the more reliable narratives on social history that these could reveal. More positively, this means that there is much scope for research and study and I hope that the book will raise the profile of that potential.

Following on from the Rock books, it was important that the contents were accessible and not just to geologists and archaeologists. There are those who have understandable reservations about increasing amounts of popular and 'grey literature' intruding into the academic realm. But some go further and talk of the damage done by 'dumbing down' scholarly research for public consumption. My personal view is that if we scientists fail to effectively share our knowledge widely then by default we only advance a myopic and untenable attitude. One that does nothing to rectify our ongoing collective inability to explain research and its implications clearly and concisely to society (who probably fund the studies!).

This also overlooks the need, when multidisciplinary science has never been more important, to translate complex topics for those researchers outside one's own discipline. While it is essential for specialists to focus on the detail of their fields, answers to the big issues are more likely to come from collaboration between disciplines. Ignoring these wider channels can only be to the detriment of science and knowledge. As W B Yeats said, we should "Think like wise men but communicate in the language of the people." You can be the judge of whether this book has managed to adhere to that.

The Whin Sill and Hadrian's Wall from Steel Rigg

Links to resources

Books

Geology of Hadrian's Wall (1997) G. A. L. Johnson

Hadrian's Wall (2000) David J. Breeze and Brian Dobson

Britain begins (2013) Barry Cunliffe

Hadrian's Wall Guidebook (2019) English Heritage, David J. Breeze

Northumberland Rocks (2021) Ian Jackson

Cumbria Rocks (2022) Ian Jackson

Durham Rocks (2023) Ian Jackson

Web pages

Open the camera app on your phone and point at the QR code

BGS Northern England Regional Guide

https://webapps.bgs.ac.uk/Memoirs/docs/B06817.html

BGS Geology Viewer

https://www.bgs.ac.uk/map-viewers/bgs-geology-viewer/

BGS GeoIndex

https://www.bgs.ac.uk/map-viewers/geoindex-onshore/

Society of Antiquaries of Newcastle upon Tyne

https://www.newcastle-antiquaries.org.uk/

Cumberland and Westmorland Antiquarian and Archaeological Society

https://cumbriapast.com/cgi-bin/cwaas/cp_main.pl

English Heritage Hadrian's Wall

https://www.english-heritage.org.uk/visit/places/hadrians-wall/

The Geological Society

https://www.geolsoc.org.uk/

The author

Ian Jackson was born in Carlisle and did his degree in geology and geography at the University of Newcastle upon Tyne. Ian is a Chartered Geologist and Fellow of the Geological Society. He spent 18 years surveying the geology of parts of the north of England for the British Geological Survey. Later he was responsible for national and international programmes that produced the first UK, European and global digital geological maps and made them web accessible.

Ian retired from the position of BGS Operations Director in 2011 and lives halfway along the Wall, in the middle of a Roman camp. He hikes somewhere in the north of England every week and is a volunteer National Trust guide at Housesteads Fort. In addition to many scientific maps, articles and reports, he is the author of three recent popular books about the rocks of Northumberland, Cumbria and Durham.

Sources and acknowledgements

The maps, memoirs and reports of the field geologists of the British Geological Survey are the prime geological sources for this book. But I would especially like to thank one of those geologists, Brian Young, for enduring endless questions about Northern England rocks and minerals.

I read hundreds of archaeological and historical books, reports and academic papers and theses as I tried to bring my knowledge of the human component of the evolution of the region up to a level to be able to write this book. Reference and bibliographic lists are not appropriate in a book like this, but I would like to acknowledge the many authors of these works.

Without conversations and advice from many archaeologists, especially Hum Welfare, Nick Hodgson, Andy Birley, Erik Graafstal, David Breeze and Mike Collins, the result would be much poorer.

Abigail Burt, a Quaternary geologist in Ontario, with a passion for archaeology, was an invaluable sounding board as I worked on the research and provided essential GIS and graphic design input. I am grateful to Abigail and Tony Cousins for proof reading and editing drafts and to Nick Hodgson and Hum Welfare for reviewing the archaeological content. But as ever, responsibility for the interpretations and any errors in this final publication is mine.

There are many others who knowingly, or unknowingly, through casual conversations and emails contributed to the work; their names are listed below – thank you. I am grateful to friends who I persuaded to hike strange routes, to what must have seemed even stranger destinations and put up with an endless stream of geoarchaeology. As usual it was my iPhone which took most of the photographs and I am grateful to David Barnes for letting me backseat-pilot his drone.

I thank Chris Hartnell of Northern Heritage Services for yet again having the faith to publish a book on stones and Sue Ward for her invaluable help with promoting it.

Finally, thank you to my wife, Gill and family for putting up with piles of books and maps everywhere and being patient when mealtime conversations regularly strayed onto rocks and Romans.

Warren Allison
Albyn Austin
Clifton Bain
Lindsay Banfield
Vanessa Banks
Mike Bishop
Sally Boardman
Mark Brennan
Liam Bullock
Richard Burt
Helen Caldwell
Brian Campbell
Louisa Campbell
Dana Challinor
Peter Claughton
Tony Cooper
Alex Croom
Jim Crow
Andrew Crowson
John Cruse
Anthony Douglas
David Dungworth
Elaine Edgar
Jane Evans
Andrew Fielding
Frank Geicco
Elisabetta Gliozzo
Martin Goldberg
Andy Hammon
Nick Henderson
Martin Henig
Jan Hennissen
Dave Heslop
Emma Hinchliffe
Doug Holliday
Andrew Howard
Mike Howe
Chris Johnson
Elizabeth La Duc
Russell Lawley
Elizabeth Laycock
Jon Lee
Dave Lowe
Angus Lunn
Jan Mather
Ollie Mather
Mike McGuire
Frances McIntosh
Tim Mighall
Bruce Napier
Don O'Meara
Andrew Parkin
Stephen Parry
Martyn Pedley
Allan Pentecost
Paul Picavet
Elizabeth Pickett
Andrew Poad
Derek Proudlock
Elsa Price
John Reid
Sibrecht Reniere
Alastair Robertson
Mike Rogerson
Ian Rotherham
Steve Rozario
Alan Rushworth
John Scott
Dorothy Sharpe
Kerry Shaw
Ruth Siddall
Mike Stephenson
Maurice Tucker
Ian Tyler
Robyn Veal
Mark Warhurst
Guy Wilson
John Zant